# CONSTRUCTION
# SAFETY
# PLANNING

# CONSTRUCTION SAFETY PLANNING

DAVID V. MacCOLLUM, P.E., C.S.P.

JOHN WILEY & SONS, INC.

New York   Chichester   Weinheim   Brisbane   Singapore   Toronto

This publication is designed to provide accurate and authoritative information in regard to the subject matter covered. It is sold with the understanding that the publisher is not engaged in rendering professional services. If professional advice or other expert assistance is required, the services of a competent professional person should be sought.

**Library of Congress Cataloging-in-Publication Data:**

MacCollum, David V., 1923-
   Construction safety planning / David V. MacCollum.
      p.   cm.
   Includes index.
   ISBN 0-471-28669-9
   1. Civil engineering—Safety measures—Planning.   2. Building—Safety measures
—Planning.   I. Title.
TA192.M33   1995                                                                94-38803
624'.028'9—dc20

10  9  8  7  6  5  4  3  2  1

# Contents

# Preface

 T his book has been written as a practical, step-by-step guide for construction safety planning from design to completion and as a ready reference for landowners who are going to build, for architects and engineers who will design a structure, for construction managers who will oversee a project, for construction superintendents who will supervise various phase of the work, and for the safety professional who, as a member of the planning team, will be responsible for seeing that safety is incorporated into all construction phases of a project. The information and recommendations in this book have been compiled from sources believed to be reliable, but I cannot guarantee or assume responsibility for their correctness, viability, or completeness, knowing that no text can cover all situations that might arise during construction and that readers must tailor such information to meet their needs and use additional sources that will suit them, as other or additional safety measures most likely will be required for particular circumstances.

The lack of a safety rationale in our social systems by which we insure against loss, mete out justice, and govern is probably the greatest hindrance to our achieving a safe workplace; and this deficiency, coupled with the frailties of human nature exhibited by management and workers alike in communicating hazard information, has created a monster that will be difficult to remedy with a single, all-encompassing stroke. Safety engineering is not the simplistic activity most people think it is, as it must consider all available options and the hazards inherent in each option. Safety planning provides us the opportunity to pursue the option that has the fewest or most easily controlled hazards. With our vast computer technology, we should be able to identify those hazards that cause the greatest harm and occur most frequently; and with our great and unsurpassed engineering capability, we should be able to design and manage the control of hazards so that they cannot endanger us. We should never be caught up in risk analysis because risk evaporates when we eliminate or control the hazard.

The amount of detail within the text is not intended to overwhelm its readers with safety requirements, voluminous paperwork, and endless planning meetings. The book is scoped to provide everyone involved with a broad overview and a ready summary of many of the things to consider in planning both large and small projects. A reasonable amount of detail is given for each phase of project development and construction, but, of course, not every conceivable hazard that might be encountered has been listed. It is hoped that the book's numerous references and list of convenient sources of information will provide food for thought and assist the readers in solving specific problems not mentioned in the text.

This book can be used by construction managers—who are on the front line and generally are first to be held accountable—as a primary guide because it shows how safety can be included in the Critical Path Method (CPM) or other computer-generated formats. Specific hazards and controls needed to avoid injury, death, property damage, and costly delays are discussed in relation to the particular phase in which they most often occur. Safety planning makes it easy to pass on pertinent hazard information to crews doing the work and creates a cooperative labor–management atmosphere, enabling a project to move along smoothly and efficiently when lives are not placed in jeopardy because of uncontrolled hazards.

Many large architectural and engineering organizations that have included safety in planning have successfully avoided the high-loss catastrophes that invariable shipwreck a project on the shoals of liability. Since 1985 many legislative bills have been proposed to reform the Occupational Safety and Health Act (OSHA) to make construction safer, and each proposal has always included a requirement for a written safety plan. Some states and various federal agencies already require construction safety plans.

This book is intended to provide an easily readable road map for developing safety plans that are appropriate for any construction project. It shows how to motivate teamwork in the examination of each phase of a project, how to identify any hazards that the team may encounter, and how to define reasonable and cost-effective measures to overcome them, with a safety professional playing an essential role as a team member. Throughout the book many of the safety concepts and requirements found in the U.S. Corps of Engineers' *Safety and Health Requirements Manual,* EM 385-1-1, have been restated or summarized for the benefit of readers unfamiliar with the manual, which is internationally recognized as a leading safety reference. It is highly recommended that copies of this manual be made available on all construction projects.

Chapter 1 discusses reasons why construction is dangerous and why it needs to be made safer. Chapters 2 through 4 consider other factors that affect construction safety. Chapters 5 and 6 give an overview of management concepts that will make construction safer. Chapter 32 shows how safety planning can be a cost-effective and profitable endeavor that ensures quality as well as a safe workplace. The rest of the book covers each facet of planning that must be done prior to the commencement of construction and discusses the high-loss hazards most commonly encountered in each phase of the work.

With this book, management has available, for any project, a ready reference of pertinent safety considerations that affect not only the landowner who is providing the money, but everyone involved, including those who eventually will use the facility when it is completed. Knowing how to include and utilize the skill of a safety engineer is one of the keys to success in safety planning and safe construction.

# Acknowledgments

Early in my career I had the good fortune to work with and learn from some distinguished safety pioneers, whose concepts were light-years ahead of their time. Jim Grady, who wrote many of the safety codes for the Oregon Industrial Accident Commission in the 1945–55 period, clearly understood that effective safety requirements were not a product of minimum consensus standards. Neil Chitwood, Chief of the Safety Branch for the Portland District of the U.S. Army Corps of Engineers when the big dams were being built on the Columbia and Upper Willamette rivers, was a prophet who introduced construction safety planning into the Corps's construction contracts in the early 1950s. His forward thinking resulted in a workplace injury-experience rate that was one-fifth that of his contemporaries who were responsible for safety programs for similar construction. Bob Jenkins, Safety Director for the entire U.S. Army Corps of Engineers from the 1940s into the 1960s, made construction safety planning a Corps requirement worldwide and was able to lower its average construction workplace injury rate to one-fourth that of other projects. Bruce Linkous, P.E., C.S.P., Chief of Safety and Occupational Health, North Pacific Division, U.S. Army Corps of Engineers, is carrying on that safety-planning tradition and has kept me abreast of current requirements of the Corps.

Over the last thirty years Harry Philo, a Detroit lawyer and safety advocate, has held construction management accountable in the courtroom for death and injury resulting from a failure to include construction safety planning in their projects, particularly where multiple contractors have been involved. I am grateful for what I have learned from him.

Most of all, I appreciate the untiring support of my wife, Nancy. She has had to put up with my anger and heartache over the past forty-two years in my battles for workplace safety, which, in the beginning, were something like shoveling sand against the tide.

# CONSTRUCTION
# SAFETY
# PLANNING

# Chapter *1*

## Construction Safety Chaos

Who is in charge of the clattering train?
The axles creak and the couplings strain;
And the pace is hot, and the points are near,
And Sleep has deadened the driver's ear;
And the signals flash through the night in vain,
For Death is in charge of the clattering train.

(Punch, circa 1880)

To some people the word "plan" is a four-letter word, and I do not disagree with them, as the world is full of Monday morning quarterbacks. The following letter sent by Lawrence A. Bullis of Phoenix to *The Arizona Republic* and reprinted in *Harper's* is a reflection of the rebellious attitude often associated with safety:

Every day some new do-gooder is trying to save us from ourselves. We have so many laws and safety commissions to ensure our safety that it seems nearly impossible to have an accident. The problem is that we need accidents, and lots of them.

Danger is nature's way of eliminating stupid people. Without safety, stupid people die in accidents.

. . . With safety, however well-intentioned it may be, we are devolving into half-witted mutants, because idiots, who by all rights should be dead, are spared from their rightful early graves and are free to breed even more imbeciles.

Let's do away with safety and improve our species. Take up smoking. Jaywalk. Play with blasting caps. Swim right after a big meal. Stick something small in your ear. Take your choice of dangerous activity and do it with gusto. Future generations will thank you.

Safety planning is a tough subject because there are so many different views and associated problems in any human endeavor, making clear answers difficult to come by. We should look at the chaos around us, not just to see the damage and improvements that have been made, but to see that there has been real progress since the building of the Hoover Dam in the 1930s, which was a big improvement over the building of the Panama Canal in the early 1900s. We must try to find and correct those forces holding us back. Progress should never be limited by our acceptance of repetitive, boring, and *dangerous* work as a reasonable source of employment. Right after World War II, I worked in logging camps in the Pacific Northwest during my summer vacations while attending Oregon State College on the G.I. Bill of Rights; and I could never understand the mentality of people striking to ban the use of chainsaws to fall and buck the large trees in the Oregon forests because those who used handsaws could not compete with that technology. Who really wants to spend a lifetime doing the dullest thing imaginable—pulling on a long handsaw—when the world is so full of more interesting (and safer) things to do!

When I entered the safety profession after finishing college, I soon recognized that I had selected a growth profession because I could see that injury and death rates could be reduced by finding those factors that allowed hazardous conditions to arise and to continue to exist when there were plenty of ways to eliminate the hazards in the first place. Hazard prevention is not the problem; the real issue is that our traditional systems allow well-known hazards to occur and do nothing to control them. Many of our catastrophes occur because we have been forced to navigate through dangerous reefs, and the traditional system has denied us the use of available charts. For the last twenty-five years construction deaths, injuries, and catastrophic damage have been subjects of political debate, newspaper headlines, and litigation. As smoke is caused by fire, construction deaths, injuries, and catastrophic damage are cause by hazards. The involvement of politicians, members of the news media, and lawyers is not something that will soon disappear but is a symptom of a very serious problem: we are killing too many people on construction projects. Our national performance with respect to construction workplace injuries and deaths does not compare favorably with that of other nations (see Table 1-1).[1]

In the United States, construction workers comprise less than 5 percent of the nation's work force but sustain approximately 20 percent of the nation's workplace deaths and injuries.[2] If the construction experience were to be deleted from the national workplace death rate, the national average would be only 7.2 rather than 9.0 deaths per 100,000 employees. The U.S. construction worker, with a death rate of 39 deaths per 100,000, is exposed to a risk of death on the job over five times greater than that of other American workers.

Over the years the news media have been quick to inform the public of construction catastrophes (see Table 1-2).

The disasters soon lose their newsworthiness on an individual basis, and their cumulative effect all over the country goes unnoticed. What is not reported is the fact that there are over 2,400 construction deaths every year, or an average of over ten deaths every workday, Monday through Friday, fifty-two weeks a year. No one really knows how many are killed on construction projects, as reporting systems fail to account for the self-employed, the uninsured, and the piecework worker. If these deaths were to be added to the estimates of the U.S. Department of Labor, National Institute of Occupational Safety and Health, National Safety Council, and National Safe Workplace Institute, the figures would be much higher. Many seriously injured workers never return to the workplace and suffer a slow, lingering death; but because they do not die immediately, their deaths are not included in construction death rates. A more realistic number of construction deaths would be three to four thousand a year.

In the mid-1960s a real political push began for construction safety, and by 1969 the Construction Safety Act was enacted for federally funded construction projects. This act required all contractors to meet a high standard of care, with specific safety requirements and specifications becoming part of the bid itself

**Table 1-1.** Construction Fatality Rate Per 100,000 Workers

| Country | Rate | Lives lost |
| --- | --- | --- |
| United States | 39 | 2,300 |
| United Kingdom | 15 | 874 |
| Greece | 16 | 943 |
| Finland | 18 | 1,058 |
| New Zealand | 18 | 1,058 |
| Spain | 24 | 1,426 |
| France | 30 | 1,771 |

**Table 1-2.** Major U.S. Construction Catastrophes, 1978–90

| Year | Lives lost | Location | Occurrence |
|------|-----------|----------|------------|
| 1978 | 51 | West Virginia | Cooling tower scaffold collapse |
| 1981 | 11 | Harbour Cay, FL | Condominium collapse |
| 1987 | 28 | Bridgeport, CT | L'Ambiance plaza lift slab collapse |
| 1988 | 3 | Milwaukee | Tunnel explosion |
| 1989 | 23 | Texas | Explosion at phillips petroleum plant renovation |
| 1989 | 5 | San Francisco | Tower crane upset during raising |
| 1990 | 0 | Washington, DC | Building excavation cave-in |

in government contracts; all contractors could project their costs for safety and include them as a bid item. Because contractors knew that they would be paid for complying with these safety requirements, they did not object to including strict safety requirements as a bid item. This law was soon incorporated into the Occupational Safety and Health Act (OSHA), not only for government contracts but for all industry; but industry representatives in general objected strenuously to having strict safety requirements included as their responsibility and found the immediate cost to them unacceptable, so OSHA's requirements were weakened by making minimum consensus standards sufficient. I believe that anything that is agreeable to everyone soon becomes the madness of the mob, slamming the door shut on any progress—in this case closing the door to having to abide by stricter safety requirements and reasonable hazard prevention measures and prohibiting the flow of life-saving safety information to the job site.

In the 1980s, the Reagan administration, while attempting to reduce bureaucratic interference and make the nation more productive and competitive, adopted two policies that caused the construction injury and death rate to rise. First, every new safety standard had to be screened by the Office of Management and Budget for its cost-effectiveness. The trouble with this policy was that this Office, when speculating upon immediate cost to the workplace, overlooked the fact that an uncontrolled hazard takes lives and costs the community many, many dollars. Second, the OSHA Review Commission adopted a pro-employer posture that weakened the enforcement of reasonable safety requirements. During the Bush administration, some progress was made by OSHA when it raised the penalties for violation of safety requirements and initiated some stricter interpretations of the consensus standards. Senate and House bills proposed in 1993 included requirements for construction safety planning and the hiring of construction safety specialists, as well as criminal penalties for violations.

The State of California in 1991 passed a law requiring every employer to institute an Injury Prevention Program. Such programs should reduce the number of deaths and injuries on the job site and give relief to another program that is greatly affected by workplace injury and death. Years ago Congress provided all disabled workers and their dependents with Social Security benefits because workers' compensation benefits were limited. This entitlement costs the taxpayer $7 billion a year, and the cost keeps going up. At the same time workers' compensation rates are skyrocketing. Even with the combined benefits of workers' compensation and Social Security, the totally disabled and their dependents and the dependents of deceased workers generally have to live at the poverty level. The resulting economic gap is the engine that drives personal injury liability litigation against the landowners, engineers, construction managers, prime contractors, equipment rental firms, and equipment manufacturers who are considered by the litigants to have fostered or condoned an unsafe workplace.

The construction industry is undergoing a very radical change. The general contractor is no longer always the major or principal employer. Often the "general" may be responsible for less than one percent of the payroll. The major labor force is comprised of the employees of many subcontractors, who individually have little or no control over the workplace's safety environment. Hazardous chaos reigns supreme when the general contractor assumes that safety is the responsibility of each subcontractor and does not oversee and coordinate safety concerns to see that an individual subcontractor's work does not create a hazard for other subcontractors. This change is causing a great need for centralized construction safety planning, in which the work of all parties is coordinated to avoid hazardous circumstances and one party is responsible for overseeing the entire project for safety.

The usual initial response to any reform movement is a strong "anti" position by lobbyists representing various concerned groups. Reforms affecting safety are no different. There are those who want to extend the immunity now given to employers by the workers' compensation laws to other parties, put caps on personal-injury liability, and weaken the already minimum consensus safety requirements, limit governmental intrusion, and compromise victims' rights rather than make an effort to eliminate or control hazards. This attitude reflects a lack of vision; it is a failure to understand hazard prevention as the most cost-effective approach to reducing injury and death in the workplace, and a failure to see construction safety planning as the best means of achieving this so that job site workers are not confronted with dangerous circumstances that they may or may not recognize or be able to avoid.

In personal injury litigation in which construction management is alleged to be negligent, the basis of this claim is, invariably, the total absence of safety planning. Because of its prevalence in today's world, litigation should not be overlooked as a very viable source of hazard information, nor should those involved in it conceal hazard information that might be used to prevent death or injury in the future. Construction managers need the assistance of safety professionals to conduct research on personal injury litigation arising from similar projects so that this hazard information can be incorporated into future planning.

Safety has an important role in initial project planning to identify hazards inherent in the location of nearby activities (such as schools, factories, or businesses that may be affected by the construction) and those that will arise during demolition and site preparation, in erection methods, and during life-cycle use. Construction safety planning addresses hazard prevention before hazards arise at the work site and is a proactive approach that will dramatically outperform reliance upon random safety inspections to discover unsafe worker acts or dangerous construction methods while work is underway.

This is not to say that engineers associated with construction projects either as designers or as construction managers are not concerned about safety; but, unfortunately, often their formal training has not included safety methodologies common to the safety community.[3] Engineers usually do not have at their finger-tips information about specific hazards and available safeguards and thus cannot include safety in planning. Everyone benefits when a safety engineer is included in construction planning activities, such as the Critical Path Method (CPM) or other computer-generated planning systems—especially other engineers whose training is in conventional engineering disciplines, who now have an unlimited opportunity to prevent high-loss circumstances from occurring on the project.

The dilemma that confronts the construction industry and those who provide engineering services and equipment for its support is that often all parties involved in a project seem to work in conflict with each other when it comes to safety. There are a number of business practices that either independently or in combination reduce the incentive to ensure a safe workplace:

1. Contracts that excuse the landowner, architect, construction management engineer, and prime contractor from involvement in safety matters relating to the project.
2. Contracts that do not include safety specifications and that invite hazardous shortcuts.
3. Workers' compensation immunities granted employers, even though they deliberately fail to provide a safe workplace.

4. Liability insurance policies that fail to require specific conditions and safety requirements as a condition of coverage.
5. The use of unsafe construction equipment that does not have available safeguards.
6. Reliance upon minimal and outmoded consensus construction standards.
7. A management hierarchy that fails to include a competent safety professional, who should be the principal advisor on all safety and/or hazard prevention matters, who is an effective advocate of humanitarian, economic, and public concerns, and who will provide the technical assistance necessary to achieve safety in design and construction activities.
8. Low-bid pricing of both design and construction, which leaves little margin for including safety. Even when labor and management have cooperated to address safety, developers and owners have bottom-lined the industry to death.[4]

A change in construction safety philosophy is overdue. Available progressive management systems and technology can provide substantial financial rewards when they are included in construction safety planning to ensure hazard prevention. To provide for sound safety planning, an injury information system should be developed that identifies hazards so that they can be anticipated; and necessary hazard prevention measures should be incorporated so that major high-loss hazardous circumstances do not arise, injury and death are reduced, costs are lowered, and governmental intrusion and unfavorable publicity are avoided.

REFERENCES

1. National Safe Workplace Institute, *Unmet Needs—Making American Work Safe for the 1990's,* Sept. 4, 1989.
2. National Safe Workplace Institute, 1990.
3. Main, Bruce W. and Frantz, J. Paul, "How design engineers address safety: What the safety community should know," *Professional Safety,* American Society of Safety Engineers, Feb. 1994.
4. Krizan, William G., Bradford, Hazel, and Setzer, Steven W., "Law of jungle is gaining strength," *Engineering News Record,* Jan. 31, 1994.

# Chapter 2

# The Role of Insurance in Construction Safety

It is understandable that children are afraid of the dark.
It is incomprehensible why men are afraid of the light.

(Plato)

Every construction manager knows about the high cost of insurance that must be paid up front before any work is done. The whole concept of retrospective rating plans for workers' compensation and personal injury liability insurance seems backwards to me in light of today's technology, which can identify hazards that repeatedly take money out of everyone's pockets and can come up with solutions that will prevent these hazards from causing further harm and monetary loss. Insurance rates should be based not upon past losses calculated in retrospective rating plans but on the savings that can be generated when safeguards and measures are taken to reduce the damaging effect of hazards that have a past record of repeatedly caused injury, death, and property damage. Construction managers should be given a choice between today's antiquated retrospective formulas and a proactive type of coverage based on projected reduced losses when hazards have been eliminated or controlled. Proactive insurance would allow premium reductions of 90 percent or more to evolve by substituting hazard prevention for the cost of paying for damages created by uncontrolled hazards.

Few think of a construction safety plan as a very cost-effective insurance policy that covers many risks and parties, from the landowner, architect/design-ers, construction management, general contractor, and subcontractors to sup-pliers of goods and services. We must abandon our conventional thinking about retrospective insurance rating plans with high premiums as a normal cost of doing business and develop a more cost-effective approach, examining where the money goes and how it can be better invested. Insurance is a very poor long-term investment when compared to an investment in eliminating or controlling hazards, which diminishes the need for insurance. Insurance is the most costly method of controlling loss caused by hazards, as it provides no method to fund their prevention. When a construction manager has prepared a thorough safety plan that includes specific requirements for hazard prevention, then insurance premiums for the project should be reduced, providing more than enough money to pay for prevention of specific high-loss hazards. Most insurance companies apportion their premium dollars by allocating one-third to sales, one-third to administration and profit, and one-third to pay for losses. What planners in their right mind would put $100 in the bank to pay for household improvements and throw two-thirds of it away? Insurance compa-nies have become cash cows because the higher the premiums are, the more money they have to invest. There is no incentive for hazard prevention because lower premiums mean a lower cash flow. The insurance company ends up having two-thirds of every dollar for itself and one-third to pay claims that should not have occurred in the first place had a good safety plan been in place. The following verse makes good sense:

It is unwise to keep a hazard, as you may have to pay for the harm it causes.
When you pay to remove the hazard, it costs you a little money—but that is all.
When you do not pay to remove the hazard, you sometimes lose everything to pay for the harm it caused.
If you keep the hazard, it is well to save something for the risk you run.
And if you do that, you will have more than enough to pay for the removal of the hazard. Now you have no risk to run!

Workers' compensation laws in the various states generally have reduced contractors' incentives as employers to provide necessary hazard prevention measures, granting them immunity from liability with no-fault insurance. In some states this extends beyond the immediate employer to the prime con-tractor, and in some jurisdictions it extends to the landowner who oversees the project.

Over eighty years ago when American society was labor-intensive, the first workers' compensation laws were drafted. These laws provided a "Band-Aid" solution for unsafe workplaces, as they compensated workers for injury and loss of time but did not hold employers accountable for removing hazards from the workplace so that injury or death could not recur. Under these laws, employers cannot be held grossly negligent for having a dangerous workplace that violates recognized safety standards. Thus they give the employer unwritten permission to ignore safety requirements. In 1983 an article was published that identified the weaknesses caused by such immunities:[1]

> Thus by failing to impose full cost of work-related accidents on employers, the workers' compensation system creates inadequate economic incentives for work place safety. . . . The workers' compensation system is thus unsatisfactory both as a means of providing adequate relief to the injured workers and as a mechanism for creating the proper incentives for the reduction of accidents and related costs.

The summary of the article suggests:

> The broader judicial recognition of the exceptions not only could pave the way for legislators to reach decisions on the individual exceptions but might encourage legislatures to pursue a comprehensive examination of workers' compensation policy required for effective and thorough reform.

A paper by James Edward Koch clearly identifies workers' compensation laws as part of the problem that creates hazardous construction work sites.[2]

Because of the immunities granted under our present workers' compensation laws, injury losses make no dent in an operating budget, as the cost of workers' compensation is considered to be part of the fixed overhead. Contractors who do provide safe work environments are penalized by such a system because their rates go up when too many injuries occur in a particular rate class, automatically subsidizing the unsafe contractor to continue maiming and killing the work force.

Workers' compensation laws have created an uneven playing field in the area of liability because the employer, a major player, is usually home free; and this has caused the injured or their dependents to look to other parties who may have contributed to the dangerous or hazardous circumstance leading to the injury or death, creating the nation's present "liability crisis." Personal injury liability cases are discussed more fully in Chapter 3. Besides deterring hazard prevention, workers' compensation laws have placed an additional burden on

Social Security, Medicare, and sometimes welfare systems, forcing taxpayers to pay for the unsafe workplace.

It is not to be inferred that the entire workers' compensation system should be trashed. Quite the contrary—it only needs a few adjustments to make it a very viable force for safety.

Having wraparound insurance programs for large projects provides planners the opportunity to develop an insurance plan that can reduce premiums and pay dividends at the completion of the project. Such a wraparound insurance program would cover all types of coverage for everyone involved on the project and give a more level playing field where everyone on the project must abide by the same set of rules, which would reduce hazards and, consequently, injuries and claims. The better the safety planning, the better the return is on the wraparound insurance program.

During planning, the total monetary outlay for workers' compensation, personal injury, third-party liability, automobile, fire, and other insurance policies should be examined very closely, as their combined cost could spell disaster, sometimes exceeding any anticipated profit from the project. Instead of continuing as usual and blindly accepting the excessive cost of premiums as a cost of doing business and the only way to go, you should consider what an investment in safety could do for your profits. It might be a more beneficial choice! Before purchasing any insurance coverage for a project, you should ask prospective insurers the following questions:

1. Who is their safety director? Request a copy of his/her resume.
2. What safety services do they plan to provide?
3. Will they provide a ten-year summary of claims and litigation experience on similar projects? In each case this summary should include:
   a. The specific hazards encountered.
   b. A brief description of the hazardous circumstance that caused the injury, death, property damage, or compensable construction delays.
   c. The name of the claimant/plaintiff and, if litigated, the court and case number.
   d. The amount paid to satisfy the claim either by voluntary agreement or as a result of litigation.
   e. The amount of legal fees paid to defend against the claim or court action.
   f. The name, address, and telephone number of the claimant/plaintiff's attorney.

With this information, you can focus on the high-cost hazards that can be avoided, including appropriate hazard prevention measures and safeguards in your planning, and you should be able to bargain effectively with your insurers for reduced premiums because of your planned investment in safety. By requesting such information, you can discover other needed information: you can find out what tangible hazard prevention technical support and hazard data the insurers plan to supply, and whether the cost of litigating previous claims has been excessive. Those who defend claims usually work on a time-and-charges basis, which sometimes adds up to sheer robbery as compared to plaintiff's attorney's fees, which are usually on a contingency basis with no cost to the injured or the families of those killed if they lose. A recent newspaper cartoon humorously showed "day one" at a law school where the professor had written on the blackboard: "It's not whether you win or lose. It's how long you drag out the case for billable hours."

Far too often project management assumes a passive role when dealing with insurers and their attorneys, which can lead to unnecessary and costly expenditures, as both insurers and defense attorneys may not be familiar with hazard prevention and may attempt to defend indefensible hazardous circumstances. A far better approach would be to retain as a consultant a responsible safety engineer, who has provided expertise in similar litigation, to give the construction manager his or her insight about whether to proceed with a trial or try to reach a settlement agreement. I have seen many cases in which the project manager was totally unaware that the hazard causing injury or death on the project had occurred many times before on other projects, and had been previously litigated with verdicts in favor of the injured. Construction management should retain authority over any litigation and make the final decision as to its disposition, and should require its insurers to report any hazardous circumstance that must be avoided in the future to prevent a recurrence. Punitive damages can be assessed when insurers and construction management deliberately ignore hazardous events that occur repeatedly. It is wise to follow up all litigation experience by taking all measures necessary to prevent a similar type of incident from happening again.

Insurance reforms are desperately needed to aid the construction industry and could reduce insurance costs as much as 80 to 90 percent. For example:

1. Insurance companies should compile data from their claims records on the hazards that cause injury and death. In this era of computer technology, information concerning the specific hazards that make construction sites unsafe should be compiled and shared with the insured. This critical hazard information should include the work methods or the construction equipment that gave rise to the hazard that caused the injury or death, as well as

the available safeguards. With this information, the construction industry would know where to look and what to do for hazard prevention. The insurance industry could be the biggest contributor to the safety information highway referred to in Chapters 3, 4, and 5.

2. Insurance companies should require that specific hazard prevention measures be in place *before* coverage is granted. Hazard prevention is profitable. The strict application of hazard prevention has an earning power that surpasses that of most investments. When those covered by Factory Mutual policies installed sprinkler systems and other fire prevention devices required by the insurer, they found their premiums reduced to 1/15 or 1/25 of the premiums for unprotected structures. Factory Mutual applies one-half of this reduced premium to hazard prevention activities, thereby shifting its principal cash flow to an investment in the lives and safety of others and creating more wealth for all.

3. Workers' compensation insurance should no longer provide immunity to an insured who deliberately fails to provide available, recognized safeguards. Such an insured also should be liable outside of workers' compensation for any injury resulting from such failure. By placing the non-complying employer in jeopardy, a strong, self-policing incentive would be created to provide a safe workplace and would soon rid the construction industry of the poor performers.

4. Insurance companies should develop accrediting programs for various types of construction activities, with primary emphasis on a certification process for prospective insureds in which they must meet specific hazard prevention conditions to qualify for a drastically reduced premium.

5. Large construction projects should have the option of a self-insured program that consolidates all risks presently covered by workers' compensation and personal injury liability into a single wraparound program so that the landowner, developer, architect, designer, and construction management have the opportunity to turn insurance expenditures into a long-term investment in hazard prevention through a safety plan encompassing both design and construction. By carrying its own insurance, construction management could penalize contractors who failed to meet the requirements of the safety plan; and such penalties should always be higher than they would have been if the contractor had provided the necessary safeguards in the first place.

6. Small construction firms with similar activities should be able to form their own insurance fund with an accrediting program in which qualifying participants comply with a standardized safety plan and program.

We need to move from risk management to hazard identification and prevention. If we do that, we will have fewer risks to manage with costly insurance.

## REFERENCES

1. "Exemptions to the exclusion remedy requirements of the workers' compensation statutes," *Harvard Law Review,* Vol. 96, pp. 1641–61, 1983.
2. Koch, James Edward, "Liability and the Injured Worker in the Construction Process," Technical Report 283, The Construction Institute, Department of Civil Engineering, Stanford University, Palo Alto, CA, June 1984.

# Chapter 3

# The Role of the Justice System in Construction Safety

The twelve jurors were all writing very busily on slates. "What are they doing?" Alice whispered to the Gryphon. "They can't have anything to put down yet, before the trial's begun." "They're putting down their names," the Gryphon whispered in reply, "for fear they should forget them before the end of the trial."

(Lewis Carroll, *Alice's Adventures in Wonderland*)

A construction manager may never give the justice system a second thought when making construction safety plans until subpoenaed for a deposition about an injury or a death that occurred on the job site, and then it is too late. It would be helpful in safety planning for the safety engineer to be able to obtain a summary of litigation arising from personal injury on similar projects so everyone would know the hazards that must be addressed in planning. Unfortunately the justice system does not work that way, and there is no safety information highway presently available.

The U.S. justice system has had a very positive, continuing influence on construction safety. The courts have been providing a public forum on construction management and the role of the landowner, architect, engineer, construction manager, prime contractor, subcontractors, suppliers, component vendors, equipment manufacturers, and all others who have an active interface or involvement with the construction process. When personal injury occurs, it often is initially assumed that the victim was at fault, the equipment was misused, or the parties overseeing the project had no safety responsibility. The courts have shown that these assumptions are not always correct and that on-the-job injury is often caused by hazards that could have been discovered if safety planning had been done.

I am not a lawyer; I am a safety engineer. The comments and opinions presented in this chapter are my personal observations, offered from a safety engineering viewpoint. The hazard information developed in personal injury litigation is of such significance that it should not be overlooked; it should be circulated so that others will know which hazards are the maimers and killers on the job site, and steps can be taken in safety planning to avoid them. Since 1969 the National Safety Council, in its *Accident Prevention Manual for Industrial Operations,* has proclaimed: "Needless destruction of life and health is a moral evil," and "Failure to take necessary precautions against predictable accidents involves moral responsibility for those accidents." I have compiled the information on litigated cases presented in this chapter strictly from a hazard prevention viewpoint; it should not be regarded as legal advice, as that must be given by a lawyer.

The public has been exposed to many opinions on personal injury litigation, from people who have labeled it a hidden tax, a deep-pockets scam, or a stifler of enterprise but never seem to mention the uncontrolled hazards that are repeatedly causing injury. Recently the *Wall Street Journal* reported on a survey conducted by Jury Verdict Research, which found that 59 percent of the time the injured prevail in their claims against makers of industrial products.[1] Little information is available on settlements, but it is generally thought that out of all the cases filed, three-quarters of them are settled and never go to trial. It also was reported that "the study looked at 1,100 out-of-court settlements and found that they averaged 28% less than jury awards," but because the sample was so small, the researchers could not determine the reliability of the data. Across the nation there is a tremendous hazard data resource that should be tapped to make construction safety planning a viable, cost-reducing program that saves lives as well. It must be understood that in most circumstances the only time individuals or corporate bodies become vulnerable to personal injury liability litigation is when they have failed to control hazards. Liability is the price that is paid for condoning unsafe and dangerous conditions and circumstances.

Two areas can be identified as the principal sources of construction-related injury: defective machinery or equipment and management that condones unsafe methods or practices. The justice system has been focusing on these two areas, but the hazard information stays buried in court records instead of being put onto a safety information highway. The construction industry's entire effort seems to be spent on defending hazardous circumstances over and over again from courtroom to courtroom and not in seeing that the hazard prevention information submitted to the court is given to the right parties so that lives can be saved in the future. Hundreds of people are injured and killed from powerline contact with cranes, and many such cases have been litigated against the same groups of defendants; but the word never seems to get back to the right people to do something about the hazard in their safety planning. For example, let us consider the monumental number of cases involving powerline contact by cranes. Many landowners, developers, construction management firms, electric utilities, crane manufacturers, and crane rental firms have been to court time after time defending themselves, yet they never seem to do anything with the vast amount of hazard data that has been produced. After a trial is over, the information is locked away forever in court records, and nothing ever is done with it to save other workers' lives.

The justice system provides monetary relief for those injured by hazardous circumstances if negligence can be established. However, the courts base their findings upon negligence rather than focusing upon the hazard. To further complicate the matter, the tracking of hazards from one court to the next is difficult because the justice system works on a case-by-case basis, and no summaries are made unless the case is appealed. These appeals offer only a sketchy description of the hazard and concentrate on the precedents of law that established negligence. The American Trial Lawyers Association (ATLA) has, for years, published a monthly *Law Reporter,* which contains summaries of cases that have been tried and, sometimes, incidentally, describes the hazardous circumstances involved, a review of which shows many instances of repetitive litigation from the same hazard. Because the justice system has no formal means or mechanism to collect and summarize hazard information developed in personal injury litigation, this vital information lies buried in court archives, and the construction industry is kept in the dark about repetitive hazards that are causing injury and deaths all over the country. The reality is even worse than this because it is a common practice for defendants to conceal hazard information deliberately by having the court issue protective orders to seal this information from public access. This practice has been outlawed in several states because it has been recognized that such orders jeopardize the lives and safety of others who can be entrapped by similar hazardous circumstances. Protective orders in themselves are in direct conflict with the universal common-law duty-to-warn, as the concealment of hazard information prevents others from hearing about a life-threatening hazard because the hazard is made a legal secret. Along the same line, it is a heinous practice of blackmail that endangers the life and safety of others for a defendant to propose an enticing settlement agreement with the injured in which the injured must agree not to disclose hazard information. Another form of unwarranted secrecy and an affront to the freedom of information is the removal of experts from the courtroom during the testimony of fact witnesses and other experts, which can conceal valuable information on inherent hazards and causes of injuries. My own courtroom experience has shown me that there is far too much discussion at the side bar, excluding the jury and experts from a free exchange of information. Consequently, judges sometimes make very erroneous rulings, as pertinent information about hazards and the safeguards available for their control are not allowed to be heard.

As discussed in Chapter 2, workers' compensation gives immunity from lawsuit to the immediate employer, even when the employer has deliberately failed to provide reasonable safeguards or to practice safe construction methods. This defective law, by excusing one party from liability, had created an uneven playing field for those attempting to rid construction sites of dangerous conditions. Because they are exempt, employers do not have to use available safeguards, putting another roadblock in the way of safety. A few states have begun to address this inequity. In Texas, if negligence on the part of an employer can be shown as the principal cause of a death, the deceased's family can recover directly from the employer. In several other states, courts are allowing defendants to bring the employer in as a third party when the employer's conduct can be shown to be negligent.

There are many good safeguards and safe practices that have not gained universal acceptance in construction planning because their effectiveness has been savagely attacked in litigation by defendants who want to avoid liability for not using them. Such bad-mouthing has created a vicious circle. It deters the use of a safeguard, thus causing more injury, more lawsuits, and more bad-mouthing over the same hazard, which could have been easily controlled by the safeguard. This senseless charade makes a mockery of the courts because more time is spent on ridiculing safeguards than in endorsing them.

Construction personal injury lawsuits have created a great deal of hard feeling in the area of hazard prevention. Rather than addressing hazard prevention to reduce the necessity for litigation, defendants have placed strong emphasis on tort reform to limit liability. We have been caught in the "House that Jack Built" syndrome:

This is the safeguard that was not used to control the hazard;
This is the hazard that caused the injury;
This is the injury that was the basis of a lawsuit;

This is the lawsuit that resulted in a big award;
This is the big award that was used as an excuse for tort reform;
This is the tort reform that encouraged negligence;
This is the negligence that arose when the safeguard was not used;
This is the safeguard that was not used to control the hazard. . . .

Construction safety planning is the key to preventing hazards, which prevent injuries, which prevent litigation. Tort reform does nothing for hazard prevention. Rather than tort reform, we need to require the justice system to participate in a safety information highway, as discussed in Chapter 4. All hazard data developed by the litigants in discovery should be made open to the public to let all available hazard information stand the light of day. The secrecy requested by defense lawyers stifles publication of hazard information and is a major cause of repetitive injuries from similar circumstances. In my opinion, lawyers who initiate protection orders, request isolation of experts, and withhold safety information are, in all reality, great hindrances to safe construction planning.

Construction managers and safety engineers, when considering possible deletion of controversial safeguards from a safety plan, should be aware that many of the old excuses used in the courtroom for not applying available hazard prevention measures are not valid anymore:

1. "Nobody else is using the particular hazard prevention device." (The courts do not give credibility to the argument that the rest of the industry does not accept its use.)
2. "The safety device is not totally and absolutely reliable." (You would not discontinue use of pneumatic tires because they blow out or puncture, or electronic navigation equipment in aircraft because it sometimes fails momentarily. The courts have held it is better to have reasonable protection most of the time than no protection at all.)
3. "The contractor did not want the safeguard installed." (If a life-threatening hazard exists that a safeguard would control, construction management has been held responsible if an injury occurs because it oversees the work of the contractor.)
4. "There was nothing in writing about this hazard." (Some courts have held that when a hazard is known by planners, the hazard and its prevention should have been included in the safety plan.)
5. "The contractor should have known better." (Courts have held that construction management should know what its contractors are doing.)
6. "The injured works for someone else, not us." (It has been held that when one contractor creates a hazard that causes another contractor's employee

to get hurt, construction management is still responsible, as it must assure that all employees have a safe place to work.)

7. "The contract gave us very little control." (Courts have held that those who have the power of control can be held responsible for its exercise.)
8. "We did not know about the hazard." (In cases where it has been proved that past records showing knowledge of the hazard were destroyed, the courts have shown little sympathy for the defendant.)
9. "We did not give any safety advice so cannot be held accountable." (If you are overseeing the work and have stop-work authority, this excuse does not get you off the hook.)
10. "We assumed the contractor was a safe, competent, and careful one because it specialized in this type of work." (The hiring of a contractor without checking its previous performance is foolhardy.)

Instant communication has made today's world one of constant change. When disaster strikes, the whole world knows about it and wants to know why it happened and what can be done to keep it from happening again. With technology advancing faster than standards and regulations can be written, a safety plan provides us the opportunity to be innovative in the use of the best available safeguards. If no standard or regulation exists to cover a specific hazard, we should do some research and see what is available in other technologies that could be incorporated into the safety plan to control the hazard. In 1937 when they built the Golden Gate Bridge in San Francisco, those in charge proposed the unheard of use of safety nets, which, up until that time, had only been used in circuses—and their use saved nineteen lives. The courts usually take into consideration those who have tried to save lives, and the best evidence construction management can have to show its care and concern for the lives and safety of all workers on a project is a thoughtfully prepared and diligently administered construction safety plan.

REFERENCE

1. Geyelin, Milo, "Product-Liability Suits Fare Worse Now," *Wall Street Journal,* July 12, 1994.

# Chapter 4

# The Role of Government in Construction Safety

A fig for those by law protected!
Liberty's glorious feast!
Courts for cowards were erected,
Churches built to please the priest.

(Robert Burns)

Government, as it interfaces with construction safety, is, in reality, twins or Gemini, with two distinct personalities, one that of the legislators (politicians) who pass laws and the other that of the caretakers (bureaucrats) who administer the law. Unfortunately, legislators cater to special interest groups, which may or may not provide the insight required to provide necessary economic incentives to ensure a safe workplace. The caretakers of the law then have imperfect laws to administer that do not create any economic incentives for safety. As discussed in Chapters 2 and 3, the employer's exclusion from liability under the workers' compensation laws makes the playing field uneven for those involved in a particular project because a major player is exempt from liability. Many think the answer is to extend the exemption to other parties; but this gives no one the incentive to work toward the common goal of developing a construction safety plan in which the landowner, developer, construction management, general contractor, subcontractors, those providing utilities, services, equipment, or materials, and anyone else who has an interest in the project work together on hazard prevention. The concept of such cooperation has already been adopted by twelve countries in the European Economic Community, where the landowner or the developer is responsible for safety planning for both design and construction.[1] However, it has been the practice in the United States for all parties to attempt, by contract, to isolate themselves from safety responsibility and to pursue their own agendas, which usually puts a very low priority on safety.

To overcome existing conflicts that prevent all parties involved in a construction project from working together on safety, government needs to develop a safety information highway that collects data from all sources. Present reporting systems address everything about an injury circumstance but the hazard that caused it and the safeguards needed to keep it from happening again. A general reorganization of injury reporting must be undertaken so that hazard and hazard prevention data will be uniform throughout all governmental agencies, workers' compensation programs, and insurance companies; and a way must be found to access the detailed hazard information and preventive measures contained in discovery and court records of personal injury liability litigation so that it, too, can be channeled into the safety information highway. Construction management needs to know which activities, processes, machines, and equipment are harboring high-risk hazards, and what safeguards will control such hazards. Presently, the government and its agencies, with a few exceptions, cannot even suggest the use of anti-two-blocking devices on cranes or ROPS on road rollers, and there seems to be no way for workers to put their practical knowledge to work on hazards and their control on a national basis. How can the construction industry improve itself if the present system keeps it in the dark? When a comprehensive summary of hazards and their cures can be made available to the construction industry and the public, voluntary acceptance and use of this information by the construction industry soon will follow, as no one wants injury or death in the workplace. Worker training would be vastly improved by such a safety information highway, as pertinent information on high-risk hazards and the known available safeguards could be included in course material.

The government relies upon consensus standards in its rule making. The tragedy of this approach is that by the time an agreed-upon standard or governmental rule is written, the graveyard is already full of people who could have been saved. The reality of today's world is that technology moves ahead so fast that by the time we get the rules in place, they are outdated. This is like today's travel—we are no longer going by train; we are going by jet airplane. Standards-making bodies cannot keep pace with our fast-moving technology. In many respects, our standards- and rule-making process is like a squirrel's cage that revolves endlessly and goes nowhere. It needs to be abandoned for a more rapid response system to prevent hazards. The Australians, in their new workplace safety laws, have taken some steps in this direction. They have declared that an inherently dangerous workplace is unlawful and requires reasonable safeguards. This type of approach would open many doors for the caretakers because they would have the authority to research available safeguards for specific hazards and could widely publicize them, giving management all known, available options to control specific hazards. This would give construction management the opportunity to include appropriate safeguards in its safety plan. The succes of this approach lies in the government's not having to endorse specific hazard prevention requirements, but using the safety information highway to publicize all known and available hazard prevention measures necessary to control specific high-risk hazards revealed by the research. This changes the role of government so that it is primarily a readily accessible safety information resource.

Current governmental safety inspection programs are ineffective because it is impossible to have enough inspectors to cover all workplaces all of the time and to be there when new hazards arise. It would be better for a qualified person with appropriate credentials and certifications to make an independent, non-governmental review of a construction project and its safety plan, using the safety information highway as a source of data on hazards and safeguards, and to sign off on it before work commences if it is found to be adequate. This individual would review the project quarterly to see if the plan were working and would write a report for construction management, evaluating the strengths and weaknesses of the project's safety performance. If life-threatening hazards were found during such a review, the reviewer would be required to enter them into the safety information highway, along with the safeguards used to control them. By using the data in the safety information highway, government would be able to focus its field staff on inspecting and investigating egregious contractors who have a disproportionate frequency of injuries, who do not use safety planning or known available safeguards, or who can be pinpointed as a source of potential disaster. The fines collected by the government for a dangerous workplace then could be used to fund the research for the safety information highway. This would develop an industry-wide emphasis on safety excellence and, in due time, would effectively and substantially reduce construction injuries and deaths as well as insurance costs, and would dry up the source of personal injury liability.

Having the safety plan and construction methods audited by an independent professional safety engineer or Certified Safety Professional (CSP) would be no different from having a Certified Professional Accountant (CPA) review a project's financial records. This would put safety on a par with financing and provide an ongoing review for safety where little exists today. Already in place to carry out such safety reviews is a safety engineering profession of some 30,000 people, some 8,000 of whom are certified safety professionals, and many of whom are registered engineers. Colleges and universities have programs that offer degrees in the field of safety, including a doctorate (Ph.D.). Several states have professional registration for safety engineers, and programs are being developed that will extend safety engineering registration to all states.

Once legislators bite the bullet and give worker safety equal footing with other business considerations, the nation's economy will benefit by having fewer injuries and deaths, which will ultimately reduce building costs. The real art of government will be to develop a system that is self-policing. Today's construction industry is in turmoil because of inadequate safety planning. When determining who is responsible for an injury or a death, all parties now tend to hire attorneys to toss the question about like a hot potato from one to the other, when they should have been hiring safety engineers in the first place to develop a safety plan to avoid injury or death. Often the biggest share of the cost of construction injury or death falls in the lap of the taxpayer through Social Security, Medicare, and welfare benefits. This is no way to run a country when the life and the safety of our construction workforce is at stake, especially when 5 percent of the workforce population is employed in construction and sustains 20 percent of all workplace injuries and deaths.

Government's role in construction safety should be:

1. To require construction safety planning for both design and construction.
2. To develop a safety information highway of injury data, hazards, and available safeguards.
3. To shift the function of preliminary and routine construction safety audits or inspections to qualified independent safety engineers or CSP's.
4. To level the playing field so that parties who deliberately avoid safety planning and the use of available safeguards are not exempt from liability in the event of injury or death.
5. To use governmental fines for an unsafe workplace to fund the safety information highway.

These broad, conceptual suggestions on how government could improve construction safety would allow workers to have input into the safety information highway; would eliminate the artificial division of safety responsibilities by landowners, developers, architects, designers, construction managers, contractors, equipment manufacturers, and suppliers of services; and would creatively address the unique hazards of the construction industry.

## REFERENCES

1. Totterdell, J. M., "Construction (design and management) regulations," *1994 Proceedings,* American Society of Safety Engineers Professional Development Conference, Las Vegas, NV, June 20–25, 1994.

*Chapter 5*

## *Bringing Order Out of Chaos*

As modern civilization is constantly creating artificial dangers to life, limb, and health, it is imperative upon us to provide new safeguards against these perils.

(Theodore Roosevelt)

The conventional method of ensuring construction safety is to rely primarily upon someone making safety inspections to assure compliance with the Occupational Safety and Health Act (OSHA). Such monitoring provides only for random sampling and is a very vulnerable method because the safety inspector may not be aware of or see all the hazardous circumstances, or may not be at the hazardous location at the right time. On many construction projects it is difficult to have enough safety-inspection manpower. For years the U.S. Army Corps of Engineers' civil districts, in order to add many "eyes" to the safety program, have required its engineers, when developing design or visiting construction sites, to include safety performance as part of their priorities and observations to assure compliance with pertinent provisions of the U.S. Army Corps of Engineers' *Safety and Health Requirements Manual,* EM 385-1-1. To maximize the participation of all the extra "eyes" for safety, it has been a leader in providing safety training to all its engineers as an initial requirement and in a continuing educational development program. The same type of management extension of safety responsibilities has been adopted by a number of design and construction firms and has provided an effective, ongoing oversight-review of hazard prevention.

A more efficient means of preventing construction injuries and deaths than watchfulness is construction safety planning, which identifies most hazardous conditions or circumstances before they arise on the construction site. This is 100 percent better than having to be at the right place at the right time to identify and control hazards by chance, or having workers cope with a hazard. However, construction safety planning does not eliminate the need for the watchful "eyes" of everyone on the job site.

Construction safety planning develops in the design engineer, project engineer, construction manager, and all other engineers—regardless of their particular discipline, be it civil, mechanical, electrical, industrial, or safety—they should have the skill of anticipating what hazards might arise. It also develops participation and teamwork for hazard prevention in these four broad functions:

1. Design.
2. Project management.
3. Phase construction activities.
4. User operation of the completed facility.

To successfully involve professionals in all aspects of construction safety, a basic understanding of the objectives and the terminology associated with hazard prevention is needed.

Many employers, employees, and others may not know that a particular work situation might be very unsafe and could kill or maim without giving them even a moment's notice. They are not aware that a hazard is an unsafe condition that exists in three modes:

1. *Dormant:* A dormant hazard is one that is created on the drawing board, from either omission or inclusion. It lies in wait, undetected, in proposed materials or in construction methods that will be used.
2. *Armed:* A dormant hazard silently awaits the right work circumstance to become armed and ready to cause harm. Sometimes a hazard has not been detected during construction planning and is armed, ready to be activated; or the interaction of equipment used on the job and those operating it or working around it may trigger a dormant hazard into the armed mode. Also a source of energy may exist that is capable, in certain circumstances, of causing injury or damage.
3. *Active:* The right combination of factors will trigger a dormant or an armed hazard into action. Energy will be released in such a manner as to cause

injury or damage. At this point it is too late to implement safeguards to avoid or prevent injury, or for the doomed to escape.[1]

Two very important aspects of the hazard must be considered: (1) the likelihood of its occurrence and (2) the severity of the consequences. Even when the likelihood of the hazard's springing into action and causing injury or death is remote, if the risk of harm that can be inflicted by the hazard is severe, the hazard must be eliminated or physical safeguards provided. Hazards can be very unforgiving. It should be understood that if a serious hazard is not recognized or perceived and avoided by every individual exposed to it, someone could be "punished" with crippling injury or death. This reality contrasts sharply with the hasty conclusions and fuzzy thinking so prevalent today that it is everyone's individual responsibility to be constantly alert for hazardous conditions or to tolerate a hazard as an acceptable risk, even when simple safeguards would prevent, disarm, or control the hazardous physical condition.

One factor that has contributed to the high proportion of injuries and deaths experienced in the construction industry is the misunderstanding many people have about the role of the safety engineer. The industry's understanding of the role of the safety engineer has been too narrowly scoped for too long. It has not understood that the safety engineer has skills that can significantly cut costs, increase productivity, and eliminate retraining of new employees to take the place of injured ones, and are its best guarantee that the profit drain from injury, death, and property damage ultimately can be reduced. The skill and the knowledge inherent in this humanitarian and profit-centered occupation have been virtually untapped. Too many people in management mistakenly believe that a safety engineer's only value to an organization is to inspect and to assure OSHA compliance. They have never understood that a safety engineer possesses the expertise to:

1. Identify hazards.
2. Evaluate the risk of harm those hazards create.
3. Inform all levels of management of the magnitude of the peril and ensure appropriate training.
4. Define safe design criteria or construction methods and prescribe appropriate physical safeguards to avoid or reduce injury and prevent property damage.[2]

Figure 5-1 shows the organizational structure that will best assure the effective functioning of the principle safety engineer where one organization is responsible for designing and building a large project. To take the best

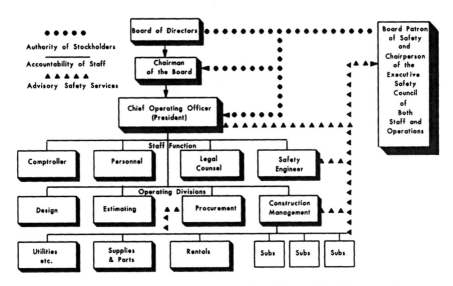

**Figure 5-1.** Safety management organization for design and build construction.

advantage of the safety engineer's training and skill, any organization engaged in construction must see that the involvement of the safety engineer with the total project is not restricted, from design through its entire life cycle, to assure safe erection methods and the safety of the facility's user. For smaller firms that either contract projects designed by other parties or act as construction managers, Figure 5-2 shows a desirable organizational structure for safety.

Do not let interpretation of such charts be so rigid that engineers who have particular expertise cannot be assigned to special tasks where their knowledge will enhance the safety of the project. Engineers should not be hard-wired into a box on a chart, but should have the flexibility to receive new assignments or promotions.

If death, maiming injuries, and property damage are to be reduced, all authority for hazard prevention must rest with the Chief Executive Officer (CEO), who is ultimately responsible for the economic well-being of the organization, must understand the economic and personal advantages of an effective hazard prevention program, and must be intimately involved with safety to see that all potential risks that can arise from uncontrolled hazards are reduced. Not only can an organization sustain a monetary loss when a hazard is activated; the CEO must understand that he/she can be sued, fired, or even put in jail because of failure to control injury, death, or property damage due to failure to assure safety planning.

The safety engineer who serves as the project safety director should be the

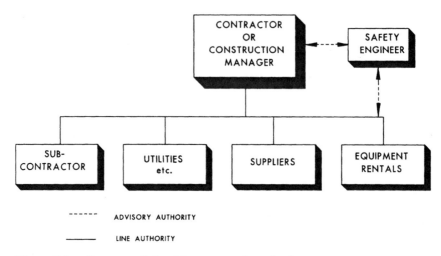

------   ADVISORY AUTHORITY

_____   LINE AUTHORITY

**Figure 5-2.**   Contractor Safety Management Organization.

principal advisor to top management on all safety matters, and must be given authority to see that safety is integrated into all levels of management. In this advisory capacity, the safety engineer must adroitly and effectively communicate to the CEO in explicit detail the dangers and the risks the organization faces. The safety engineer should be trained to tell it like it is and to communicate the magnitude of a hazard, its specific danger, and the nature of the peril to the CEO. Once the danger and its economic consequences are made known, a discerning CEO will immediately realize the worth of the safety engineer and become his or her greatest advocate and supporter. A safety engineer must have support from the top, or any hazard prevention program, no matter how life-saving and cost-effective it is, will never have the clout it needs for dealing with the various echelons of management and their internal pecking orders. The key to the safety engineer's success is to coordinate the technical assistance of all engineers in the conventional disciplines so that design options can be integrated into the project to meet specific safety criteria. It is the role of the safety engineer to develop a common objective to control hazards at the time of design, rather than to assume that perfect human performance will avoid the hazards. Perfect human performance can never meet the test as a reliable safeguard because those who are confronted with hazardous circumstances may not recognize their danger and respond in the proper manner.

   Another untapped resource that will enable the construction industry to make giant strides in its hazard prevention program is the gathering of meaningful injury and damage data to show where trouble spots lie. Even in this age of magnificent computerized data processing equipment, there are few data bases that gather definitive data on specific hazards. Death and injury may have occurred from the same hazard source over and over again, but no one has bothered to computerize data on these repetitious events to smoke out the causative factor—a dormant or armed hazard that became active. Unsafe equipment or machine design might be a contributing factor; so data should also be collected about the equipment, giving the type, make, and model. Equipment design that has not been subjected to an intense system-safety hazard analysis is often the cause of death and injury, as a dormant design hazard can lie in wait for the first person to make the fatal move that will activate it. If the hazard is not eliminated after the first occurrence, it can go on to kill and maim again. We cannot always blame the person for being injured, as a dormant hazard that the person was not even aware of might have caused the person to respond in an unsafe manner.

The term "accident" is a misnomer, as it implies that an occurrence was due to chance, was unforeseeable, and could not be prevented. The insurance industry is very vocal about insurance costs being raised by product and personal injury litigation, but it has failed to computerize hazard information and to pass this information onto its insureds so that known and repetitious hazards can be eliminated from construction projects.[3] Such data would not only allow the construction industry to gain ground in its hazard prevention programs, but would reduce the number of claims that it must pay because of construction injury, death, or property damage. This information would give everyone the edge on controlling the high number of hazards that are allowed to go unchecked day after day. If there were public access to insurance claim data and to litigation data so often hidden by protective orders, that would give the construction industry valuable insights into hazardous equipment or work circumstances that have repeatedly caused predictable injuries and deaths. After a hazard has been identified, the focus should be on using this information in safety planning.

Once the construction industry begins to utilize the hazard identification skills of the safety engineer, it will find that this resource is indispensable in specifying what needs to be done to eliminate hazards.

The following list of five ways to reduce the risk created by a hazard is called an "Order or Precedence," as it presents each method in order of its priority and effectiveness. Sometimes several of these preventive measures are needed to control a single hazard.

1. *Design to eliminate or minimize the hazard.* The major effort through the design phase must be to select appropriate safety features to eliminate the hazard, render it "fail-safe," or provide redundancy through use of backup

components to avoid a failure mode that would cause a dormant hazard to become active.

2. *Guard the hazard.* Hazards that cannot be totally eliminated by design must be reduced to an acceptable level of risk through the use of appropriate safety devices that will guard, isolate, or otherwise render them effectively inert. A hazard should be made inaccessible, and if this cannot be done, then nearby personnel should be protected from it. For example, management should provide guards to give physical protection against moving parts and provide line covers for energized electrical conductors to prevent powerline contact. Some other methods of guarding include:
   a. Install screens or covers over moving parts.
   b. Provide rollover protection that guards the operator in the event of upset by enclosing him/her in a protective frame.[4]
   c. Insulate the hazard so that unintentional physical contact with energized conductors does not prove injurious or fatal. This form of guarding includes insulated links for cranes and line covers for energized powerlines.
   d. Install fences, guardrails, and so on, that restrict entry.
   e. Install interlocks to deactivate the system when guards or covers are removed.
   f. Install presence sensors to give an alert when someone intrudes into a danger zone. These include infrared motion detectors, high frequency sound, radar, and light barriers.
   g. Install pressure sensors to give an alert when someone steps on a mat or drives over a cord or a hose.
   h. Make the location inaccessible.
   i. Install fail-safe systems.
   j. Provide redundancy by having a number of safeguards that must be overcome before injury can occur.
   k. Internally monitor the electrical/mechanical system.
   l. Put up staging and scaffolding, which prevent falls from elevated work areas.

3. *Give a warning.* When a hazard cannot be eliminated by applying either the first or the second method, an active/intercessory warning device should be installed that detects a hazard and emits a timely, audible, and/or visual warning signal so that people can avoid it. Examples are alarms, horns, and flashing lights. Standardization of such warning systems is imperative so that the meaning of the warning is never misunderstood. Some hazard detection systems can be wired to allow movement away from a hazardous position to a safer one or to stop any unsafe movement immediately upon detection.[5]

Signs and labels are passive warnings. They must be very explicit and state what the hazard is, what harm will result, and how to avoid the hazard. The signs for life-threatening hazards should be as pictorial as possible, with the word "DANGER" written in white letters on an oval red background with a black border. Warning systems are not substitutes for the first and second hazard prevention methods. Rather, warnings are best used to alert users to the need to be aware of a specific change of circumstances or of a dormant hazard that cannot be totally eliminated or reduced, and to tell them why the specified safeguard must be used.

4. *Provide special precedures and training.* When a hazard cannot be eliminated or its risk reduced by any of the first three methods, then management must ensure that special operating procedures, training, and audits are developed to guarantee that a viable, continuing regimen is established and maintained to avoid the hazard. Remember that training can never be a substitute for hazard elimination when life-threatening hazards are present.

5. *Provide personal protective equipment.* The use of gloves, hard hats, safety shoes, aprons, goggles, safety glasses, life lines, life jackets, and other protective equipment at all times also will protect users from injury. Remember that this last method only guards against injury, and the hazard is ever present if not eliminated or reduced by the first two methods. This hazard prevention method also requires user acceptance. Bulky, ill-fitting, cumbersome, inconvenient, and restrictive personal protective equipment often is discarded and not used.

In dealing with high-risk, critical hazards, a complete safety network of physical safeguards must be established to achieve an acceptable level of hazard reduction, and, to reach this goal, one or more of the five methods of hazard prevention outlined above must be employed. The effectiveness of a specific hazard control device should be judged on its ability to serve as part of a "team" of controls. No safeguard should be discarded simply because it cannot be totally reliable under all circumstances. To reliably assure the lives and safety of others, a system of redundancy must be created through the use of a variety of safety design features that minimize the hazard, provide physical barriers, give warning, or require special procedures so that they will collectively prevent the hazard from becoming armed or activated. It is not prudent to substitute a lower Order of Precedence measure for a higher one. A warning should not be used when guarding would control the hazard.

Construction management needs the expertise of the safety engineer to continuously gather and interpret data from all available sources. Only by keeping abreast of failure and injury data from all sources possible can an organization hope to cut its losses from these sources and begin to earn profits from the safety effort. Some other tasks that the safety engineer can perform are:

1. Conduct research on occurrences and statistics, using a variety of sources on injury and/or death occurrences on construction projects.
2. Conduct literature searches for supporting data and pertinent information on hazards.
3. Review applicable standards (both national and international) for the prevention of hazards, and interpret them for management.
4. Review records such as patents and commercial products to ascertain the state-of-the-art hazard controls available.
5. Prepare a system safety hazard analysis (see Figure 5-3). Such analysis includes consideration of four interrelated variables:
   a. Human performance/propensity for human error.
   b. Construction equipment performance.
   c. The interface of human performance with construction equipment.
   d. The environment affecting the construction equipment.[6]
6. Provide both design criteria and management systems to achieve a safe construction environment and ensure their inclusion and implementation.

The State of California recently enacted a requirement for all employers to have an Injury Prevention Plan to make safety a functional part of all workplaces, including construction. This requires:

1. Identification of a specific person or persons with authority and responsibility for implementing the program.
2. Policies that ensure that employees comply with safe and healthy work practices (training or disciplinary procedures for employees who fail to observe safe practices).
3. Policies for communicating hazards to employees.
4. Procedures for identifying and evaluating workplace hazards, including the scheduling of periodic inspections.
5. Procedures for investigating occupational injury or illness.
6. Methods and procedures for correcting unsafe conditions and practices.
7. Training and instruction.

Construction safety planning brings order out of chaos. It is an ongoing activity that begins when the first concepts of a project are formed, before

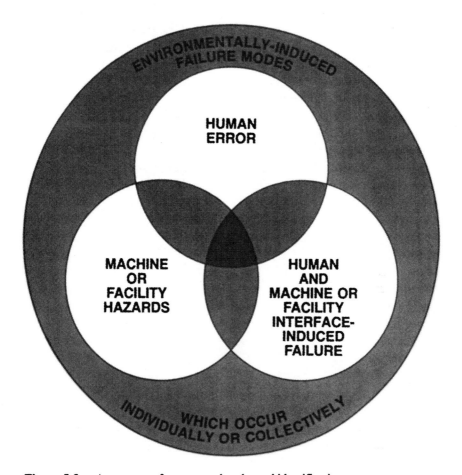

**Figure 5-3.**   A systems safety approach to hazard identification.

design is initiated, and is continually reexamined and expanded in greater detail as the project evolves from the time of design to the safe use of the completed facility.

REFERENCES

1. MacCollum, David V., "Foreword," *Readings in Hazard Control and Hazard Materials,* American Society of Safety Engineers, Des Plaines, IL, 1985.
2. MacCollum, David V., "Time for change in construction safety," *Professional Safety,* American Society of Safety Engineers, Des Plaines, IL, Feb. 1990.

3. MacCollum, David V., "Safety: Are we making progress?" *Professional Safety,* American Society of Safety Engineers, Des Plaines, IL, Feb. 1978.
4. MacCollum, David V., "Rollover Protective Systems (ROPS)," *Automotive Engineering and Litigation,* Vol. 2, Chap. 1, Peters and Peters, Editors, Wiley Law Publishing, Colorado Springs, CO, 1988.
5. MacCollum, David V., "Crane design hazard analysis," *Automotive Engineering and Litigation,* Vol. 1, Chap. 8, Peters and Peters, Editors, Wiley Law Publishing, Colorado Springs, CO, 1984.
6. Vincoli, Jeffrey W., *Basic Guide to System Safety,* Van Nostrand Reinhold, New York, 1993.

# Chapter 6

## *Project Safety Planning Concepts*

We, the people of the United States, in order to form a more perfect union, establish justice, insure domestic tranquility, provide for the common defense, promote the general welfare, and secure the blessings of liberty to ourselves and our posterity, do ordain and establish this Constitution for the United States of America.

(Preamble, The Constitution of the United States of America)

T o meet the requirement set forth by our forefathers to promote the general welfare, safety planning has three basic mandates:

1. Safe design.
2. Safe erection.
3. Safe use.

The design of airports, bridges, canals, dams, electric transmission lines, factories, harbors, highways, mines, office buildings, power plants, railroads, refineries, smelters, and tunnels will all have an impact on the environment. As the public has become more and more active in promoting the protection of our environment, it has become aware of the needless death and injury that arise during construction and is demanding that something be done about this. The answer is to include safety in all planning as an active participatory function, not as a management ritual performed by rote, as intelligent planning gives people an opportunity to find new ways to eliminate hazardous circumstances by bypassing old methods and habits.

A mostly overlooked advantage of having a safety engineer participate in planning for design, erection, and use is cost-effectiveness. Too often the financier, architect, engineer, manager, and owner do not find safety important until construction is under way or a catastrophe has occurred. In this context, safety engineering is like trying to find the plugged outlet drain in the bottom of a muddy pond when the pond is full of alligators. Including safety in design is cheaper than negotiating change orders with contractors or rebuilding a project after it is in use.[1]

Safety planning must be incorporated within the Critical Path Method (CPM) network as a part of project scheduling so that hazard prevention requirements can be adddressed before dangerous circumstances arise and costly interruptions occur. A good example is the delivery of heavy items that must be lifted by a crane. If delivery is scheduled when the items can be directly unloaded into place, redundant and dangerous handling, storage, and transport are avoided. Safety planning of this nature eliminates the use of a crane to unload engines, turbines, transformers, machines, air-conditioning units, or other pieces of heavy equipment at a remote storage yard where they can be damaged by weather or other activities before being relifted, transported, and unloaded for installation. The more times that a single piece of equipment is handled, moved, stored, worked on, or altered, the more opportunity there is for hazardous circumstances to arise. Handling items only once is safer and less costly and can be programmed into the CPM network or other computer-generated planning system.

The safety engineer can provide detailed summaries of the accident experience of similar facilities, the hazards to be avoided, and how to control them effectively, which is very valuable when one is inputting data into any planning network. This type of information binds the financier, architect, engineer, manager, owner, and safety engineer into a team.

Figure 6-1 shows the hazard prevention triangle. One leg is cost, and the second is productivity/utility, with safety requirements as the base of the triangle. This triangle represents the team's design objectives. The productivity or the utility of the completed facility is dependent upon safety. Failures traced to preventable hazards increase operating costs.

In contemplating a new project, the following initial considerations should be addressed:

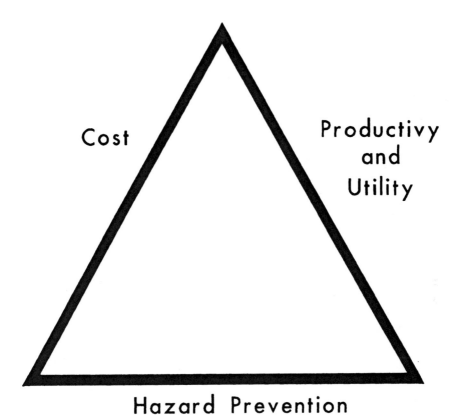

**Figure 6-1.**

1. Nature of the terrain:
   a. Soil conditions.
   b. Slope.
   c. Vegetation.
   d. Drainage.
   e. Contamination (toxic substances).
   f. Buried utilities and/or facilities.
2. The local environment:
   a. Surrounding neighborhood.
   b. Weather, climate, and prevailing winds.
   c. Aboveground utilities and easements.
   d. Access roads and streets.
   e. Aircraft flight patterns.

3. Political considerations:
   a. Building restrictions and codes.
   b. Ethnic relationships.
   c. Community activities and goals.

Hazard prevention needs to be incorporated into all initial concepts of every construction project with the same priority given to environmental concerns and costs. When hazard prevention is given the same priority as cost, utility, and productivity in the conceptual planning stage, it sets meaningful priorities for the design team. Adverse media coverage, political action, and litigation only arise after a hazardous circumstance raises its ugly head or a disaster occurs.

Safety planning in the conceptual stage is the first stepping-stone on the way to a positive, proactive, hazard prevention mind-set that creates an ongoing priority for safety through each stage of development and construction. Safety planning does not need to create voluminous paperwork; it only needs an outline of anticipated hazards incorporated into the general planning. In the early stages it only needs to be included in the environmental impact statement.

## REFERENCES

1. Brauer, Roger L., *Safety and Health for Engineers,* Van Nostrand Reinhold, New York, 1990.

# Chapter 7

## Safety and the Design Team

". . . but there's one great advantage in it, that one's memory works both ways."

"I'm sure mine only works one way," Alice remarked. "I can't remember things before they happen."

(Lewis Carroll, *Through the Looking-Glass*)

To many, avoiding a disaster before it happens is like stepping through a looking glass; they think it cannot be done. But safety planning can make it happen ! Everyone wants a project to proceed during construction without the stress of death or injury, and wants the completed facility to have no hazards to plague the user. Quality design and construction make a project profitable for everyone.

In its day, the design of the *Titanic*, which divided the hull into several compartments with watertight bulkheads, was a great advance in safe ship design. However, when the vessel collided with a submerged iceberg and the hull ripped open almost from bow to stern, it became apparent that this design was no match for such a hazard. Would the use of a double hull along with the compartments, as found in warship design and now adopted in oil tanker design, have saved the *Titanic*? Did the captain of the *Titanic* really understand the limitations of the hull design if the ship were to strike an iceberg at full speed ahead? With today's vast computer technology we can step through the looking glass to provide design teams with criteria based on hazard data and specify appropriate safeguards.

New design concepts often are rejected because they appear to be dangerous. A hazard analysis made at the time of design can, in most instances, reveal available hazard prevention measures. The acceptance of a radically new type of design by the U.S. Army Corps of Engineers for the navigation locks at the Dalles Dam in Oregon in the mid-1950s is an outstanding example of how incorporation of reasonable hazard prevention measures specified in the hazard analysis overcame all objections. Conventional design called for excavating sufficient space on each side of the lock to construct gravity-type concrete walls, and then backfilling behind them (Figure 7-1).

The lock would lift or lower river barges almost 110 feet as they proceeded through it, requiring that an excavation of 180 feet be carved in brittle, easily fractured basalt rock formed by layers of successive lava flows millions of years earlier. Instead of using conventional lock construction, the new design called for a six-foot-thick concrete plaster wall rock-bolted to the hard lava. Objections to this design centered around the excavation, which required drilling and shooting and would constantly expose workers to falling rocks from the excavated walls because of vibration and the effects of freezing and thawing in winter and expansion in summer. Initially, overhead barricades or roofs above

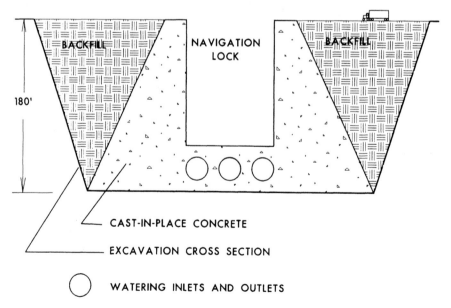

**Figure 7-1.**  Navigation lock, conventional design.

the drillers were considered, but the feasibility of heavy-enough portable sheds was questioned by design engineers and construction managers. The Portland District staff safety engineers were consulted, and they introduced a method using chainlink fencing that could be rolled down the side of the excavated rock wall like a window shade as the excavation progressed (Figure 7-2).

This simple innovative safeguard provided protection from rock falls during excavation until the concrete wall had been cast in place and proved to be exceedingly cost-effective, saving several million dollars by giving those bidding on the project a specified safeguard on which to base their estimates— which, incidentally also allowed them to reduce their estimated drilling costs by over 50 percent. The inclusion of safety engineers on the design team will not only foster new design concepts and the benefits of value engineering, but

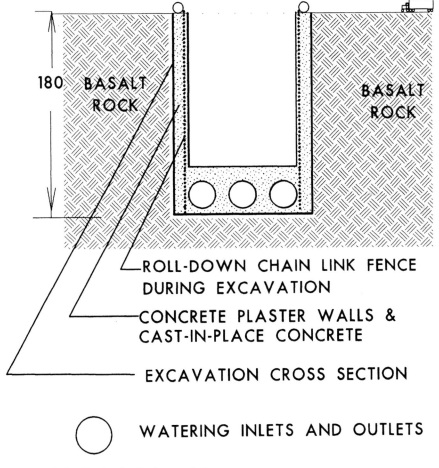

**Figure 7.2**  Navigation lock, new design.

will enhance the quality of the workplace by making it safer. When excavations are contemplated, specific requirements for shoring, which should be in compliance with OSHA, should be included at the time of design. An excellent reference is the book *Excavation Safety.*[1]

The design of a project often is spread among many different parties, each responsible for a particular segment, with little overall review to determine whether, when assembled together, the segments will create any serious hazards during construction. For example, a worker fell twenty feet to his death during construction of an office complex as he was welding the hangers of a precast concrete cladding panel to the building's steel framework. Investigation revealed that the design of the hangers was such that the panels could not have been welded into place from inside the building where the worker would have had effective fall prevention by standing on the floor, eliminating the need for him to be suspended outside the building to make the welds. The hangers were anchored in the panels in such a way that they would not hang plumb, requiring the crane from which they were suspended for installation to exert a sidepull so the hangers could be properly attached to the building frame. Thus the panels were defective because attachment to the building had to be made from the outside, and they would not hang plumb when being lifted into place by the crane. Had a safety engineer familiar with construction practices been part of the design team, the defective hanger attachment system would have been identified and corrected.

Many large firms have combined architectural and construction activities into a single firm to "design and build," making a very close relationship between design and erection processes. One reason why the Corps of Engineers has had a significantly low injury rate in its construction is that it coordinates the design with erection procedures. When the architect and construction management are separate entities, planning needs to include safety coordination between the two. Close coordination should be developed with the designer of the structural steel so that appropriate holes (1/2") can be specified as anchor points for safety lines; this provides for the safety of steel erectors by making it easy for ironworkers to install I-bolts, which make excellent attachment points for overhead safety lines to which the ironworkers can attach their safety belt lanyards or fall protection devices. When necessary, the attachment of uprights for railings is made easier by putting simple bolt holes on the horizontal beams. The project safety engineer and the design team should plan for easy attachment of fall protection systems, including safety nets. Nigel Ellis's book, *Introduction to Fall Protection,*[2] gives excellent guidance on how to design for fall protection.

Safety planning during design also must consider the durability of cladding materials and the type of glass to be used to avoid the problems that arose in Boston where some of the glass sheathing fell off a building, making it necessary to replace the whole thing. The design of connecting systems for both precast concrete and cut-stone facing is critical. A registered professional engineer who is knowledgeable in connecting systems should design and supervise this work. In some circumstances, cladding systems are subcontracted to a design group, a fabricator, and an installer, who may work independently and need to be under the supervision of an engineer to assure that the pieces are assembled and can be installed safely. Building cladding should be designed so that it can connected to the structure from inside the building to avoid working on suspended scaffolds. Provision should be made in the design for the safe cleaning of windows on the outside of high-rise buildings, as well as for inner atria, open entranceways, and so on.

A system review is a vital process that ensures that the design creates no hazards in construction and ultimate use of the facility. By the end of World War II, a system safety analysis was found necessary in the development of military aircraft. By 1963, system safety had been documented in military specifications (current identification is MIL STD 882), and it also has been adopted by many organizations outside defense and aerospace establishments.[3] Several large "design-and-build" architect/construction firms now have system safety staffs. Designers wishing to become familiar with system safety concepts may refer to the *Basic Guide to System Safety* by Jeffrey W. Vincoli.[4]

The design team should be responsible for coordination of recognized construction methods so that unnecessary hazards are not created. Often design choices determine construction methods; so the safety engineer on the design team must understand construction. As a guide for the design team, the safety engineer should prepare a written hazard prevention plan for each project that:

1. Establishes and defines a safety management structure to develop the meaningful participation of the entire design staff in a hazard prevention program. The scope and the function of the safety engineer should be set forth in this plan.
2. Defines pertinent reliable safety requirements, including applicable consensus standards of the American National Standards Institute (ANSI), American Society for Testing and Materials (ASTM), and Society of Automotive Engineers (SAE), as well as military standards and OSHA or other standards such as those of the U.S. Army Corps of Engineers' *Safety and Health Requirements Manual,* EM 385-1-1.
3. Lists potential high-risk hazards, such as falls from elevations, excavations, powerlines, underground utilities, soil conditions, toxic materials,

wastes and/or contaminants, fumes, fire, common work areas (particularly where production must continue while new building additions are being made), public vehicular and rail traffic through the construction area, and any other hazards inherent in the project.

4. Lists design concepts governing the construction methods to be used, such as tilt-up, lift-slab, steel-frame, reinforced concrete, prefab concrete beams, decks, and so on, so that special conditions to ensure safe construction methods can be included in contract specifications.

5. Requires that the entire design team receive appropriate hazard prevention training.

6. Requires hazardous site and construction conditions to be identified and taken into consideration in construction scheduling.

7. Requires preparation of construction specifications that identify applicable safety standards and any special safety conditions necessary to control high-risk hazards.

8. Requires preparation of change orders to overcome specific hazards that arise during construction.

9. Requires review of shop drawings to make sure all contractors' installations are safe.

10. Requires maintenance of a record of the safety plan, actions, reports, and minutes.

Critical to the development of a written safety plan is the retention of a competent safety engineer who holds a Professional Engineer (P.E.) license, preferably in the safety engineering discipline, who is a Certified Safety Professional (C.S.P.), and who is experienced in construction methods and system safety. The safety engineer's duties should include:

1. Being a member of the design team and the principal advisor to the supervising agency, landowner, developer, and architect on each project on all matters that concern the safety of (1) all those who are associated with or will be engaged in construction activities, (2) the public who will be exposed to construction activities, and (3) those who will use the completed facility.

2. Providing tutorial technical safety assistance to the landowner, developer, architect, engineer, and design group and identifying and arranging necessary safety training for them, as their training and experience most likely will not be in safety but in a conventional discipline.[5] Such assistance will include safety expertise on hazard identifcation and prevention and the use of system safety techniques, including human factors, reliability, fall

prevention, fire protection, and maintainability, to assure that all designs incorporate reasonable safeguards to avoid personal injury, property damage, and/or delays during the entire life cycle of the facility through construction to demolition.

3. Conducting safety research on hazards inherent in the project being designed, while using past occurrence data as a means of pinpointing specific hazards that create accident-producing circumstances. Sources of hazard information are the National Safety Council, U.S. Department of Labor, National Institute of Occupational Safety and Health, American Association of General Contractors, National Safe Workplace Institute, and other bodies involved in hazard prevention. A wealth of information also can be gathered from litigation involving high-loss construction injury and damage claims. Data on available safeguards that will prevent such accidents can be obtained from these sources. A reference library to support and document safety research about hazards identified on the particular project should be established and maintained.

4. Reviewing and signing off on all plans, specifications, and change orders by name and date of review, identifying hazards that will need to be addressed during construction, and defining appropriate preventive measures that must be incorporated into design. A written log of identified hazards, recommended preventive measures, and actions taken must be maintained.

5. Reviewing contractors' shop drawings to identify hazardous components, identifying appropriate hazard prevention measures to eliminate any hazards, and maintaining a written log of identified hazards, recommended preventive measures, and actions taken.

6. Securing appropriate safety references for the personal use of the landowner, developer, architect, engineer, design group, and project inspectors, including basic safety requirements as set forth in OSHA, the U.S. Army Corps of Engineers' *Safety and Health Requirements Manual* EM 385-1-1, consensus standards, and other pertinent safety requirements. Where special safety conditions are listed in the contract to overcome specific hazards, a separate memorandum should be prepared summarizing them for easy reference, and a copy given to each party.

7. Establishing a simplified procedure for reporting and recording hazardous conditions and practices observed by the landowner, developer, architect, engineer, design group, and its inspectors in the normal course of site visits during construction so that these observations can be forwarded to construction management for follow-up.

8. Maintaining, as a matter of public record for the life of the project, copies of all documents relating to safety, including: the Hazard Prevention Plan; special contract provisions relating to hazard prevention; record of hazards

and corrective action taken; safety changes in plans, specifications, and contractors' shop drawings; safety deficiencies found by the architect, engineer, or design personnel during site visits; and all death, injury, and property damage records. All records should be retained for at least five years after completion of the project. Retention of these records prevents misinformation from being used as a basis for liability claims.

With a safety engineer on the design team, designers of the various components of the project will have safety expertise, research, and assistance available in the developmental stage, which avoids the reworking of drawings and specifications, or, worse yet, being confronted with having to rebuild a portion of a facility found to be dangerous. There is no substitute for one-on-one, face-to-face assistance to each member of the design team on an as-needed basis. The safety engineer has an overview of the entire project, and from that vantage point is able to see if a component is unsafe, if a combination of components will be hazardous, if any part of the facility will be dangerous to erect, or if any part of the completed facility will be dangerous to operate or maintain as designed.

Additional information about incorporating safety planning into design can be found in Chapter 30, "Facility Planning and Design," of *Safety and Health for Engineers.*[6]

## REFERENCES

1. Suprenant, B. A. and Basham, K. D., *Excavation Safety,* The Aberdeen Group, Addison, IL, 1993.
2. Ellis, Nigel, *Introduction to Fall Protection,* 2nd Edition, American Society of Safety Engineers, Des Plaines, IL, 1994.
3. U.S. Department of Defense, MIL STD 882, *System Safety Program Requirement.*
4. Vincoli, Jeffrey W., *Basic Guide to System Safety,* Van Nostrand Reinhold, New York, 1993.
5. Main, Bruce W. and Frantz, J. Paul, "How design engineers address safety: What the safety community should know," *Professional Safety,* American Society of Safety Engineers, Feb. 1994.
6. Brauer, Roger L., *Safety and Health for Engineers,* Van Nostrand Reinhold, New York, 1990.

# Chapter 8

## Contract Safety Requirements

A fence at the top of a cliff is better than an ambulance in the valley below.

(Anonymous)

As soon as the design is completed, contract specifications should be prepared to accompany the design plans so that the prospective contractor will have a road map to follow in planning for the job. Just as you would not use an outdated road map when planning an automobile trip across the continent, you should not rely upon old contract specifications as a guide in writing specifications, as they were probably written by a lawyer who knew nothing about safety or hazard prevention,[1] and they could be filled with meaningless boilerplate terms. With respect to project safety, contract specifications need to establish specific guidelines to control all anticipated hazards by:

1. Naming the person who will be responsible for overseeing contractors' performance and seeing that it is being done in a safe manner. This individual must have stop-work and stop-payment authority to hold the general contractor responsible for a safe workplace, including the activities of all subcontractors and suppliers of services and materials.
2. Requiring all contractors to prepare and submit an acceptable project hazard prevention plan that defines supervisory and employee safety training prior to the commencement of their particular work at the job site.

3. Listing specific published safety standards and hazard prevention requirements with which all parties must comply.
4. Listing special hazard prevention requirements that are to be followed for unique hazards not adequately defined in provisions referenced in item (3) above.
5. Listing qualifying requirements for eligible contractors to ensure that bidders are restricted to those whose past performance shows that they are careful, competent, and safe contractors.

Many standard-form agreements between the owner and the contractor usually are consistent with formats developed by the American Institute of Architects, as they give the owner the right to stop work and do not hold the architect responsible for the control or change of construction means, methods, techniques, sequences, or procedures, or for safety procedures and programs carried out in connection with the work. Such wording relieves the architect of responsibility for minding the store when it involves the life and the safety of the workers and others on the job site. The way to overcome this gap in defining safety responsibility is for the landowner to retain a construction manager or place all of the work under the direct supervision of one general contractor. When safety responsibility is divided among many parties without any clear centralized authority, safety chaos is sure to follow.

No wonder construction death and injury rates are so high—construction contracts are defective when they do not address safety responsibility in specific terms. Many plans and specifications go into great detail about the type of water faucets, beams, pipes, bolts, bricks, and other materials to be used on the project and call for physical inspection to ensure compliance. If contractors cannot be trusted to provide good-quality materials, why do we trust them with the lives and the safety of the most precious ingredient in construction—the people who erect and assemble the structure—and make no mention of any safety requirements? The most cost-effective time to provide for hazard prevention is in the plans and the specifications of the contract.

In recent years a number of architects and/or engineers have been held negligent in personal injury suits when it has been found that they were aware of the danger and had the necessary stop-work and stop-payment authority but made no effort to assure that the contractor took appropriate preventive measures for life-threatening hazards. These decisions were based on the fact that safety is considered a nondelegable duty. When a landowner's construction manager has the staff assistance of the architect and a safety engineer and has direct control over the general contractor, and the contract specifically defines safety responsibility and requirements, the work can be safely accomplished if the construction manager has the gumption to exercise personal authority to ensure the fulfillment of all safety provisions.

A study conducted by members of the Civil Engineering Department of Stanford University and written by Levitt, Parker, and Samuelson states: "Owners who actively involve themselves in selection and monitoring of contractors with respect to construction safety performance have significantly lower accident rates on their construction projects."[2] The U.S. Army Corps of Engineers, on construction projects supervised by its civil districts, has for nearly fifty years actively overseen the safety performance of its construction contractors and has had an injury and death experience rate one-fourth to one-fifth that of its contemporaries engaged in similar construction activities. The controlling factor has always been prompt, iron-fisted enforcement of contract safety provisions. Many contractors working under Corps contracts have profited from safely managed and hazard-free construction projects.

The key to a safe construction project is to require all bidders to submit a written safety plan with their bids, as outlined in Chapter 11. The solicitation for bids must include guidelines for an acceptable safety plan and state clearly that the substance of the safety plan will be reviewed, and its adequacy will be a determining factor in who shall be selected as the contractor. It is recommended that bid solicitations require contractors to include in written safety plans the requirement that the U.S. Army Corps of Engineers' *Safety and Health Requirements Manual,* EM 385-1-1, will be furnished as a working reference for construction superintendents and foremen as a day-to-day guide. This manual is available from the Superintendent of Documents, U.S. Government Printing Office, Washington, DC 20402, for approximately $11.00. Merely referring to OSHA is unacceptable, as these standards are minimal and will not achieve a high degree of performance because they do not relate to many specific hazardous circumstances that arise on construction projects. A written safety plan enables the construction manager and the safety engineer to evaluate the bidders' concepts of safety and hazard prevention and to discover deficiencies in the proposed control of specific hazards. The bid requirement for a written safety plan puts all bidders on a level playing field; so a low bid that does not include an acceptable written safety plan will not qualify.

When special hazards are anticipated, having specific methods for their control set forth in a written safety plan enhances the contract, as each bidder is given the same basis for estimating costs. This also eliminates a lot of argument on the job site as to what is and is not an adequate control for a hazard. In the 1950s the Portland District of the Corps of Engineers included a special condition in its contracts calling for rollover protection canopies on all tractors based upon a study by its safety branch. This was ten years before the Society of Automotive Engineers published standards for rollover protective structures (ROPS).[3] The inclusion of this special condition was questioned by many because no safety standards existed at the time for the hazard of tractor rollover, but in a two-year time period four lives were saved by using this protection. Today, another area worthy of many special contract conditions involves the use of cranes at the work site. My book *Crane Hazards and Their Prevention* provides insight into this subject and lists necessary crane safety requirements and other references.[4] All equipment expected to be used on a project should be analyzed for hazards and the special conditions required to overcome them. (See Chapter 14.)

Personnel requirements also should be included in bid solicitations to assure competent employees, as outlined in Chapter 17.

Criteria concerning what constitutes a safe and careful contractor must be spelled out in the contract to limit prospective bidders to those that can qualify as safe and careful contractors. Inclusion of these criteria allows contractors to decide whether they are qualified to bid. (See Chapter 9.)

A very dangerous and widely held idea is that a total hands-off policy on safety is the way to avoid trouble. Many managers have pursued this philosophy and have caused their employers to become victims of the very thing they were trying to avoid. Some landowners are selecting iron-clad "design-and-build" turn-key arrangements where they have no on-site, stop-work authority or acceptance of materials and only evaluate the completed project, which generally removes them from day-to-day involvement with safety on the job site.

## REFERENCES

1. MacCollum, David V., "Time for change in construction safety," *Professional Safety,* American Society of Safety Engineers, Feb. 1990.
2. Levitt, R. E., Parker, H. W., and Samuelson, N. M., *Improving Construction Safety Performance: The User's Role,* Technical Report No. 260, The Business Roundtable, Aug. 1981.
3. MacCollum, David V., "Rollover Protective Systems (ROPS)," *Automotive Engineering and Litigation,* Vol. 2, Chap. 1, Peters and Peters, Editors, Wiley Law Publishing, Colorado Springs, CO, 1988.
4. MacCollum, David V., *Crane Hazards and Their Prevention,* American Society of Safety Engineers, Des Plaines, IL, 1993.

# Chapter 9

## *Qualifying Safe and Careful Contractors*

The ultimate test of management is performance.

(David V. MacCollum)

The most effective way to assure a safe construction project is for the construction manager and the safety engineer to prepare a bidders' list of contractors who are known to be competent and perform their work in a safe manner.[1] This avoids the voluminous circulation of bid proposals to those whose past safety performance would not qualify them for bidding on the project. Any contractor wanting to be placed on this bidders' list must first meet the same basic safety performance standards as contractors whose names are already on the list.

Before such a bidders' list is prepared, the safety engineer should develop strict criteria for safety performance that include the following:

1. The contractor's previous injury and damage experience. Do not be satisfied with merely examining the contractor's OSHA incident rate. Require a list of all injuries for the last five years, showing the following information for each one:
   a. Hazard.
   b. Injury.
   c. Time lost.

    d. Medical and compensation cost.

    e. Brief description of the occurrence.

    f. Hazardous equipment or facility involved in the occurrence.

    g. Hazard prevention measures taken.

2. The contractor's Experience Modification Rate (EMR) for workers' compensation.[2]

3. The contractor's written safety policy and program.

4. References from others who have evaluated the contractor's performance and were overseers of the contractor's safety program.

5. Records of the contractor's previous OSHA citations available under the Freedom of Information Act and any litigation for personal injury or property damage.

6. Whenever possible, information from the contractor as to usual construction practices, methods, and equipment used.

There is no substitute for visiting a current construction project of each candidate on the list to see if the work actually is being conducted safely.

A preliminary review such as this narrows the competing contractors to a select group who have a good safety track-record. This process also eliminates the "suitcase" contractor who bids a project with a conglomeration of low-bid subcontractors who are indifferent to safety and whose construction projects are ongoing and chaotic disaster areas. The elimination of sending out voluminous numbers of proposals to bidders whose safety performance is unknown is very cost-effective, as it avoids hours of work spent in sorting through the bids of individuals whose safety experience is unknown.

This is a very short chapter, but it describes one of the most important steps in planning. Consider your list of contractors as a rose garden, see how they bloom, and then pick only the brightest and hardiest of your team of winners.

## REFERENCES

1. Levitt, Raymond E., Parker, Henry W., and Samuelson, Nancy M., *Improving Construction Safety Performance: The User's Role,* American Society of Civil Engineers, Preprint 82-040, Spring Convention, Las Vegas, NV, 1982.

2. *Dictionary of Terms Used in the Safety Profession,* 3rd Edition, American Society of Safety Engineers, Des Plaines, IL, 1988.

# Chapter 10

## Prebid Safety Conference

Mutual understanding is the basis of cooperation.

(David V. MacCollum)

$A$fter the proposals have been forwarded to those on the bidders' list of safe and careful contractors, the next step is to hold a prebid safety conference to give bidders the opportunity to review the specifics of all safety requirements and special safety conditions and to discuss their interpretation, as well as a chance for a one-on-one review with the safety engineer to learn what must be included in their safety plans (Chapter 11). This conference sometimes results in a two-way exchange of ideas, as a prospective contractor may propose an alternative method of construction that could turn out to be less hazardous than the original plan. Some thirty-five years ago during construction of the Dalles Dam on the Columbia River, the Corps of Engineers—which was administering the project as landowner, designer, and construction manager, and was among the first organizations to hold prebid safety conferences—discovered that some of the bidding contractors had well-thought-out alternative construction methods that were safer than their own and provided substantial savings. Being able to discuss safety before the submission of bids promotes a spirit of cooperation and flexibility, making any changes for hazard prevention easier to accomplish.

Prebid safety conferences make possible a partnership for safety among those involved in contract administration and actual construction activities. All parties suffer financially when someone is killed or injured or when property damage occurs. This type of conference gives contractors a chance to question contract provisions that do not give them enough latitude to achieve the safety objectives set forth in the contract; and because everyone is given an opportunity to identify hazardous circumstances and find solutions long before work commences, it avoids complications during construction. When amendments to the contract are found necessary, they can be added so that all bidders have the opportunity to include needed safeguards in their bids.[1] You should not overlook the value of such a conference, as it provides a great opportunity to make sure that all bidders clearly understand what needs to be done to assure a safe project in terms of administrative safety management and actual physical requirements needed to prevent hazardous circumstances.

It is interesting to note that a full-page advertisement appeared in the October 3, 1994 issue of *Engineering News Record* announcing the Fifth Annual Contractors' Briefing for the Commonwealth of Massachusetts' Central Artery/Tunnel Projection, with free admission but requiring reservations. More and more often, this type of conference is becoming an accepted form of letting bidders know the intent of contract safety requirements, Such briefings provide contractors with a wealth of information about upcoming contracts and are ideal places at which to review very advantageously the details of safety requirements and specific safeguards that must be provided for equipment, as well as the reasons for such requirements.

Although this chapter is short, the worth of the prebid safety conference should never be overlooked.

## REFERENCE

1. Ely, Merrel D., "Coordinating the safety program with independent contractors," *Proceedings of the National Safety Congress,* National Safety Council, 1960.

# Chapter 11

# The Project Construction Safety Plan

If you do not think about the future, you cannot have one.
However, the truth of the matter is that each of us will have a future and we may get the future we deserve.

(John Galsworthy, *Swan Song*)

The construction manager of a project may be an individual retained by the construction management firm, a representative of the landowner or the developer, or a representative of the general or the prime contractor. This individual must have supervision and authority over the general contractor, subcontractors, all parties providing materials and services, and all other public or private activities that must be accommodated within the scope of the construction project. Safety is a nondelegable duty. It requires ongoing supervision of each subordinate level of management to ensure that reasonable hazard prevention measures are taken.

The construction manager's safety planning duties should include, as a minimum, the following:

1. Preparation and publication of a Project Construction Safety Plan that:
    a. Establishes and defines a fast-track chain of command for immediate resolution of all hazard prevention issues that concern the safety of workers or other issues that have been identified or raised by anyone associated with the project, either on or off the project.

b. Defines the individual hazard prevention responsibilities of each member of the construction management team; the safety training each must complete; the working relationship of each with the landowner, developer, architect, engineer, and design group; and the scope and the authority of the safety engineer's duties and functions (see item 13 in this list).

c. Specifies OSHA requirements, the U.S. Army Corps of Engineers' *Safety and Health Requirements Manual* EM 385-1-1, American National Standards Institute (ANSI) standards, and other specific and pertinent standards that will be relied upon.

d. Identifies high-risk hazards such as powerline-contact electrocution, excavation cave-in, confined spaces, toxic substances, and so forth, and necessary control measures; and ensures that hazardous conditions or circumstances identified by the architect, engineer, or landowner are addressed and appropriate hazard prevention measures taken (see Chapter 7).

e. Specifies the types of current certificates of inspection to be produced prior to commencement of operations or use, for cranes, elevators and lifts, pressure vessels, boats, aircraft, and other equipment; specifies necessary monitoring equipment/systems for the detection of toxic or explosive vapors, dusts, liquids, and so on; and assures maintenance of an ongoing log of these certificates, including expiration and renewing dates, and the recordings of the monitoring systems.

f. Lists equipment, machines, and tools to be used on the project to ensure that their inherent hazards are identified and safeguards provided, using the following basic format (see Chapter 14):
  (1) Type of equipment—tractors, backhoes, cranes, and so on.
  (2) Manufacturer.
  (3) Model.
  (4) Year of manufacture.
  (5) Inherent hazards and necessary safeguards, such as those shown in Table 11-1 (which is presented for illustrative purposes and is not a complete listing).
  (6) Required equipment-operator licenses, certificates, physical requirements, training, and other appropriate qualifications for drivers, crane operators, special equipment operators, boat operators, and so forth.

g. Lists and defines construction methods that will be used for each major phase of work and describes the process and safety procedures to be incorporated.

**Table 11-1.** Selected List of Hazards and Safeguards

| Equipment | Hazard | Control |
|---|---|---|
| Tractors | Rollover | ROPS (four-post) |
| | Falling objects | FOPS |
| | Backing | Reverse signal alarm |
| Cranes | Two-blocking | Anti-two-block device |
| | Overload | Load moment device |
| Trucks | Backing | Reverse signal alarm |
| | Access | Safe access |
| Compactors | Rollover | ROPS |
| Graders | Backing | Reverse signal alarm |
| | Rollover | ROPS |

h. Assures that the Critical Path Method (CPM) and other documents incorporate specific safeguards, such as deenergizing, removing, or relocating powerlines before cranes are brought onto the construction site, using shoring as trenches are excavated, installing safety lines for fall protection as steel erection proceeds, and so on; and assures that other types of planning schedules are prepared that clearly define when special safety requirements are needed so that necessary safeguards will be programmed and not overlooked.[1-4]

i. Provides for the following administrative hazard prevention documents:

(1) Plans for the layout of temporary construction buildings and facilities, including how the contractor will ensure safe use.

(2) Plans for traffic control and marking of hazards for haul roads, highway intersections, railroads, utilities, bridges, restricted areas, and so on.

(3) Details of fall prevention and protection.[5] (See Chapter 15.)

(4) Shoring systems for excavations and tunneling.[6-8]

(5) Site plans, including those for temporary power distribution systems designed to avoid work areas used by cranes or other equipment that can be raised into powerlines.

(6) Phase plans that include safe clearance procedures and permits for all high-hazard activities. (See Chapter 21.)

(7) Plans for hazard prevention surveys of the job site by competent persons.

(8) Plans for fire protection and action to take in the event of a fire.

(9) Plans for maintenance, cleanup, and safe access and egress.

(10) Contingency plans for flooding, blizzards, tornadoes, and other severe weather conditions and description of an office-trailer anchoring system to avoid windstorm hazards.

(11) Emergency plan for obtaining medical assistance, ambulances, and directions for rescue operations.

(12) Procedures for accident investigations.

2. Review of plans and specifications to identify the location and the nature of potential hazards such as powerlines, underground utilities, soil conditions, toxic wastes, common work areas, and other hazardous activities. (See Chapter 7.)

3. Review of specifications to identify appropriate safety standards and special safety conditions.

4. Establishment of criteria for qualifying bidders (contractors, subcontractors, suppliers of equipment and services, etc.) who can perform the work safely.

5. Conducting a prebid safety conference to acquaint qualified prospective bidders with all safety requirements to prevent misunderstandings and bring to their attention any proposed construction methods that would be unsafe.

6. Awarding contracts to contractors, subcontractors, and suppliers of equipment and services who will conduct their work in the safest manner.

7. Conducting a pre-notice-to-proceed safety conference for all contractors, subcontractors, and suppliers of equipment and services who will bring personnel onto the construction project.

8. Requiring construction superintendents and/or foremen to prepare a written safety plan for each major phase of the work, including hazard analyses for high-risk hazards. (See Chapter 21.)

9. Seeing that change orders are issued if required to overcome specific hazards that arise during construction.

10. Reviewing contractors' shop drawings to make sure the design is safe; and when a second opinion is needed, forwarding them to the architect, engineer, or design group.

11. Assuring that the entire construction management team receives appropriate safety training.

12. Insisting upon immediate reporting of all injuries, deaths, and/or property damage; and at the commencement of the project, appointing on a standby basis an independent board of competent professionals, who have no organizational biases or agendas, to investigate fatalities, serious injuries, and major property damage. The key to success of such a board is to appoint

those individuals who are free to respond on short notice before a site is disturbed, and who can present the board's findings promptly to the construction manager.

13. Retaining a safety engineer, who not only is a licensed Professional Engineer (P.E.), preferably in the safety engineering discipline but is also a Certified Safety Professional (C.S.P.) and who is experienced in system safety, to perform the following duties:

   a. Be the principal advisor to the construction manager on all matters of safety during construction and during the foreseeable use of the facility, including prevention of hazards that could result in personal injury, property damage, or delays. This includes the safety of all personnel on the construction site, the public who may be endangered by the construction or affected by hazards created by the operation of the completed facility, and the ultimate users of the property.

   b. Provide tutorial technical safety assistance for the entire construction management staff on hazard identification and prevention, using system safety techniques that include human factors, reliability, maintainability, quality control, environmental considerations, and fire protection as they relate to safety.

   c. Conduct safety research into hazards inherent in the types of projects under construction, giving consideration in both methods of construction and types of equipment to be used; using past occurrence data as a means of pinpointing specific hazards that create injury-producing circumstances; relying on hazard information from the National Safety Council, U.S. Department of Labor, National Institute of Occupational Safety and Health, American Association of General Contractors, National Safe Workplace Institute, and other bodies involved in hazard prevention. A wealth of information also can be gathered from litigation involving high-loss construction death, injury, or property damage. Data on available safeguards that would prevent such occurrences can be obtained from these sources.

   d. Establish and maintain a library of supporting applicable references, such as the U.S. Army Corps of Engineers' *Safety and Health Requirements Manual* and standards of the American National Standards Institute (ANSI), American Society for Testing and Materials (ASTM), and Society of Automotive Engineers (SAE), as well as military specifications and other data to support and document safety research on hazards of the particular project; and make these references available to the staff.

e. Develop and conduct pertinent safety training, or use current packaged programs, for all persons involved in the project.

f. Implement a fast-track procedure for reporting and control of hazards identified by anyone on the project. (See item 1.a above.)

g. Review and sign off on the construction plans for safety and make appropriate notations where high-risk hazards exist.

h. Review contract specifications to make sure appropriate safety standards and special conditions are included. Where reference to pertinent safety standards has been omitted, advise the construction manager of the omission and the appropriate action to be taken.

i. Provide the construction manager with supporting information for the bidders' list so that only safe and careful bidders are selected (as outlined in item 4 above and in Chapter 9).

j. Provide the construction manager with an outline of topics to be discussed at the prebid safety conference, attend this meeting to provide his/her expertise, and be sure minutes are kept and each attendee is provided a copy.

k. Assist the construction manager in the selection of a safe and careful contractor.

l. Assist the construction manager in development of the CPM to be sure the schedule indicates where high-risk hazards will occur and appropriate hazard prevention measures inserted.

m. Examine delivery schedules for heavy components that will require lifting by a crane to see that delivery is scheduled for the time of installation, to avoid unnecessary unloading, storage, reloading, transporting, and unloading.

n. Provide the construction manager with an outline of topics to be discussed with the successful bidder at the pre-notice-to-proceed safety conference, attend this meeting to provide his/her safety expertise, and be sure minutes are kept and each attendee is provided a copy. Topics should include, as a minimum, the following:

(1) Review of the contractor's construction safety plan.

(2) Discussion of potential high-risk hazards such as powerline contacts and electrocutions, confined spaces, excavations, toxic substances, and so on.

(3) Review of certificates of inspection to make sure they are current before allowing the operation of cranes, manlifts, pressure vessels, boats, aircraft, and so on, and they contain the name of the person who will maintain a log to assure that timely renewals are made.

(4) Fast-track action to be taken when on-site hazards are identified and how they will be controlled.

(5) Needed hazard prevention surveys.

(6) Anticipated hazardous construction methods.

(7) Equipment hazards (as outlined in item 1.f.5 above and in Chapter 14).

(8) Equipment tests to be conducted.

(9) Selection and training of those who operate and work around equipment.

(10) Fall prevention and protection.

(11) Shoring systems for excavations and tunneling. (See Chapter 23.)

(12) Shop drawings for scaffolds, ramps, and other temporary items.

(13) Site plan for placement of temporary construction offices and buildings.

(14) Plans for traffic control, including haul roads, highway intersections, railroads, utilities, bridges, restricted areas, and where cranes will be placed.

(15) Plans for maintaining the job site—trash control, housekeeping, safe pedestrian access and egress, and so on.

(16) Plans for fire protection and dealing with fires. (See Chapter 34.)

(17) Description and sketch of temporary power distribution system.

(18) Description of safe clearance procedures.

(19) Emergency plans for flooding, blizzards, tornadoes, and other severe weather, including a description of anchoring systems for any trailers used as offices.

(20) Emergency plans for notifying medical assistance, ambulance, and direction of rescue operations.

(21) Injury and damage reporting procedures.

o. Provide technical assistance to construction superintendents and/or foremen in preparing their hazard analyses and safety plans for high-risk activities. (See Chapter 21.)

p. Review and sign change orders to ensure that appropriate safety requirements are included.

q. Review and sign shop drawings to ensure that appropriate safety requirements are included and that they have been prepared by a competent person as required by OSHA.

r. Provide the construction manager with topics for all construction management conferences and reviews, and attend such meetings to provide technical assistance and to be sure that records are maintained of all hazards reviewed and safety actions taken.

s. Provide superintendents and foremen with timely toolbox safety topics, with enough explicit and easily understandable information that workers will have a thorough understanding of the specific high-risk hazards they will encounter during the week and the specific safety requirements they will follow.

t. Conduct frequent and regularly scheduled hazard prevention surveys to provide an ongoing overview and to make sure there is continuing effective safety participation by the entire construction management staff. An effective method for you to gain the participation of the construction management staff is to invite different members to accompany your walk-through reviews of the project, providing a one-on-one ongoing safety orientation.

u. Regularly communicate with local OSHA representatives to keep current on OSHA's requirements and priorities.

v. Investigate all injuries, deaths, property damage, and near-misses that arise during construction activities, a duty that includes:

   (1) Being an advisor to the investigating board convened by the construction manager for fatalities and major property damage.

   (2) Preparing all necessary injury reports required by OSHA, state safety agencies, and insurers.

   (3) Entering and maintaining on a computer the following information on all injuries, fatalities, and property damage, regardless of severity, to provide a data resource for trends and to show where additional safety planning is needed:

      (a) The hazard.

      (b) The host/facility, with all pertinent information.

      (c) The type of equipment, model, and serial number.

      (d) The manufacturer.

      (e) The date of manufacture.

      (f) A brief description of the occurrence, including the activity being performed, such as concrete forming, craning, earth moving, excavating, general construction, landscaping, paving, roofing, steel erection, trenching, tunneling, and so forth.

      (g) The hazard control needed.

      (h) Name, addres, and telephone number of the injured.

      (i) Names, addresses, and telephone numbers of witnesses.

      (j) Name, address, and telephone number of foreman overseeing the work at the time of occurrence.

w. Maintain a file of all safety plans, activity hazard analyses, correspondence, minutes, logs, injury data, board and other reports, computer listings, and so on, suitable for placing in a project repository to be kept for five years after the completion of the project. Retention of these records prevents misinformation from being used as a basis for liability claims.

Many projects have been complete disasters because no safety plan was developed. This chapter provides an overview of the basic and minimum items to be included in a construction safety plan. As each project has unique safety needs that go beyond the basic ones outlined in this chapter, the content of each safety plan will differ according to specific safety requirements that may have to be added. Teamwork is what counts in making a safety plan, and the construction manager and the safety engineer not only must work closely together in its formulation but must include the thoughts of the whole construction management team.

## REFERENCES

1. Monaghan, John O., "The Critical Path Method: How to use it," *Construction Methods and Equipment,* Reprint.
2. Fondahl, John W., *A Non-computer Approach to the Critical Path Method for the Construction Industry,* Technical Report No. 9, U.S. Navy, 1962.
3. Cherry, Benjamin T., "The 'SPATS' path to construction safety," *The Journal,* Sept. 1965.
4. Milor, James R., "How we apply 'Spats' in safety programming," *1969 Proceedings of the National Safety Congress,* National Safety Council, 1969.
5. Ellis, J. Nigel, *Introduction to Fall Protection,* 2nd Edition, American Society of Safety Engineers, Des Plaines, IL, 1994.
6. Cass, *Common Sense in the Common Trench,* Equipment Guide-Book Company, Palo Alto, CA, 1979.
7. Suprenant, B. A. and Basham, K. D., *Excavation Safety,* American Society of Safety Engineers, Des Plaines, IL, 1993.
8. MacCollum, David V., "Systems safety and tunnel support," *National Safety News,* Dec. 1976.

Chapter *12*

## *Selecting the Contractor for the Job*

Success to the strongest; who are always, at last, the wisest and best.
(Ralph Waldo Emerson)

Low bids are not always as cost-effective as they might seem. When a multitude of low-bid subcontractors work for a landowner in the construction of an industrial facility and death or serious injury occurs, the low bid can come home to roost in the pocketbook of the managing landowner, the construction management firm, or the prime contractor. Liability costs for deaths and injuries can far exceed the rejected higher bids of safe and careful contractors, making the actual cost of the project outrageous.

When you work with a group of bidders who have already qualified because of past safety performance and their safety plans for the project (see Chapter 9), the selection of the most qualified contractor is not a difficult task. The following guidelines also will help you in this selection:

1. Make a thorough review of each bidder's written construction safety plan, as this will quickly show which one can best construct the project in a safe and careful manner, free from hazards. (See Chapter 11.) If, despite safety plan deficiencies, a bidder seems best for the job when all other consider-

ations are weighed, these deficiencies can be addressed at the pre-notice-to-proceed safety conference and appropriate amendments made to upgrade safety performance. (See Chapter 13.)

2. Before the final selection is made, refer to each bidder's previous injury experience to determine if it is current and if any areas need improvement because of an excessive number of one type of injury from the same hazard. Check the current experience modification rating with each bidder's insurance carrier. Check with OSHA to see if there have been any recent citations of the bidder and what corrective measures are required. (This information is available from OSHA under the Freedom of Information Act.) Make an inquiry through an attorney to determine if any personal injury litigation has been filed against a bidder. If any is found, evaluate the circumstances that led to the death or the injury, the hazards involved, and the outcome. A surprising amount of safety information is available in the transcripts of the testimony of witnesses and experts. This often overlooked source of hazard data can provide you very valuable insight into the bidder's ability to conduct work safely. If this information is under protective order, be very wary of a bidder's safety performance because such orders sometimes indicate that there is something to hide, and they might be covering up a poor safety record. If the bidder has a questionable injury-experience record but still seems best for the job, the unsatisfactory injury experience must be discussed at the pre-notice-to-proceed safety conference, and specific requirements must be defined for improving the bidder's safety performance. (See Chapter 13.)

3. Make a site visit to current projects on which bidders are working to see firsthand the day-to-day quality and functioning of each bidder's construction safety program. Such a site visit will show whether the safety plan as submitted is a functioning one or just a paper illusion. A visit to the safety engineer's office also is a good idea, to see the contractor's safety reference library. A safety office that does not have a comprehensive reference library of pertinent safety textbooks, references, and standards must be considered suspect.

Examining these three areas should provide you with enough information to rate each particular bidder on a scale of one to ten, ten being the best choice, and will give you excellent insight into what items to include in the pre-notice-to-proceed safety conference. For additional information, see reference 1.

In selecting a contractor or a subcontractor for a specific job, you should make every effort to pick a real winner where safety is concerned to protect your pocketbook from possible third-party claims.

## REFERENCE

1. Samuelson and Levitt, "Owner's guidelines for selecting safe contractors," American Society of Civil Engineers (ASCE) National Spring Convention, Las Vegas, NV, Apr. 26, 1982.

# Chapter *13*

## *Pre-Notice-to-Proceed Safety Conference*

Safety performance is directly related to management involvement in the site safety program.

(Richard Hislop)

The pre-notice-to-proceed safety conference is a key meeting that will set the tone and establish the very posture of priority for safety during the life of the contract. The mandatory attendance of all parties involved is one of the most important requirements for the success of this meeting. Besides the construction manager and the project safety engineer, the following individuals must attend:

1. Landowner or party who retains ultimate stop-work control.
2. Architect and designers who have authority over design.
3. Superintendents who will oversee the work.
4. Subcontractors' managers.
5. Managers of suppliers of rental equipment.
6. Representatives of utility companies (electric, water, sewer, gas, telephone, etc.) whose lines are either on or adjacent to the construction site.
7. Representatives of government agencies who have jurisdiction, such as local law enforcement where traffic control and permits are required for alternate roadways and so forth.

It always creates good public relations to invite OSHA and/or state safety representatives, as well as safety representatives of casualty insurance companies who underwrite workers' compensation and personal injury liability coverage.

The construction manager should conduct this meeting so that all attendees have an opportunity to state their safety concerns. Many people are inclined to be passive, noncommittal, and silent at such a meeting. It is wise to query attendees individually when concerns relating to their areas of expertise or responsibility arise, to stimulate them to voice their opinions. Be sure that no time constraints are placed on this meeting, as valuable and constructive safety input might never have a chance to surface.

Be sure that complete and thorough minutes are taken to record what specific hazard prevention measures will be taken; who will perform them; when, where, and how they will be done; and who will monitor them. It is very important that the minutes also record all additions, changes, and other agreements reached, including:

1. Contract safety requirements.
2. The implementation of special safety conditions.
3. The written Project Construction Safety Plan (Chapter 11).
4. The hierarchy of authority.
5. Any other safety concerns.
6. The individuals who will be responsible for the control of specific hazards, and who must be competent and qualified in terms of work experience, hold appropriate credentials, and have authority to take immediate action if a life-threatening hazard is discovered.
7. How hazards and necessary prevention measures will be incorporated into CPM or other scheduling.

At this meeting the safety engineer will function as an advisor who has immediate resource informaton available to support the adoption of reasonable safeguards, making it easier to reach mutual agreement on hazard control. The safety engineer will be responsible for preparing discussion topics for this meeting, and the meeting's success depends upon this list's being as comprehensive as possible. (See duties of the safety engineer as listed in Chapter 11, item 13.n.) Therefore, the safety engineer should prepare for this meeting by making a full review of all hazards with a high-loss potential and by researching injury data, standards, literature, and preferred hazard prevention measures so that he/she can comment authoritatively in any discussion that may arise, as construction managers cannot always have such information at their fingertips.

The pre-notice-to-proceed safety conference lays the groundwork for developing a mutually supportive partnership on safety among construction management, the contractor, subcontractors, and all parties involved. Since 1977 the U.S. Army Corps of Engineers has made it a mandatory requirement on all civil-works construction projects worldwide for the contracting officer (construction manager) to meet the contractor and the subcontractors, before the start of every project and at each major phase of work, to discuss and analyze the hazards that may be encountered and define appropriate control measures. The North Pacific Division of the Corps initiated this program in the 1950s. This meeting also gives management the last major opportunity to disqualify a blatantly uncooperative, anti-safety contractor before becoming stuck with a real loser. The majority of contractors know safety's value and want to overcome any deficiencies in their proposed safety plans, but sometimes they need to be shown how to do this. Thus this conference can be a key factor in establishing a project partnership for safety to reduce construction hazards.[1]


## REFERENCE

1. U.S. Army Corps of Engineers, *Safety and Health Requirements Manual,* EM 385-1-1.

# Chapter 14

# Examination of Equipment, Machines, and Tools Brought onto the Job Site

Brakeless cars are not the best—even if carefully made.

(Anonymous)

When the general contractor, subcontractors, and equipment rental firms bring equipment onto the job site, it is important to have a plan to inspect each piece of equipment as it arrives, before use, to verify that it is in safe condition. The equipment must be safe for its proposed use and comply with all special safety provisions in the contract, with all requirements agreed upon at the pre-notice-to-proceed safety conference, and with all other applicable regulatory requirements and standards. Available state-of-the-art safeguards not yet incorporated into published safety requirements or standards also should be considered. To assist in such inspection, a rudimentary list should be developed for evaluating the safety of all equipment, machines, and tools, which references the source of the principal regulations, standards, and hazard prevention concepts upon which the list is based.

Equipment, machines, and tools used on construction projects can be categorized as follows, and are discussed below in this order:

1. Transporters.
2. Earthmovers and excavators.
3. Lifting equipment.

4. Scaffolding and temporary support structures.
5. Power sources.
6. Material processing.
7. Marine equipment.
8. Hand tools.
9. Miscellaneous—not elsewhere classified (NEC).

## TRANSPORTERS

Collision, upset, and backing of vehicles account for one-fourth of construction injuries and deaths. Basic requirements for safe methods of transportation can be found in the Code of Federal Regulations, Title 49, and additional details about vehicles can be found in current Society of Automotive Engineers (SAE) handbooks, particularly Volume 4, *On-Highway Vehicles and Off-Highway Machinery.*[1] Transporters have been separated into seven basic groups: trucks, forklifts, conveyors, pipelines, railways, roadways, and aircraft.

### Trucks

Trucks and other motor vehicles should be equipped with seat belts and shoulder harnesses. These devices have prevented many deaths by restraining occupants within the vehicle, and their use should be mandatory for all drivers and passengers, with severe penalties for nonuse. Airbags are proving to be very effective injury deterrents in passenger vehicles, and the construction industry should be demanding equal protection.

A new hazard has arisen with the use of two-way radios and cellular telephones. Drivers are setting themselves up for a collision when they take their eyes off the road to dial or take both hands off the wheel at the same time to hold the receiver and dial. Locating radios and telephones so that they cannot be dialed while the vehicle is in motion would avoid many collisions. A new technology of voice actuation allows drivers to keep both hands on the wheel and their eyes on the road while communicating, thus reducing this hazard.

Backup alarms have proved their worth in reducing the number of injuries and deaths associated with the backing of vehicles. With the advent of closed-circuit TV, a driver's performance, especially when backing large vehicles, can be greatly enhanced by using a rear-mounted camera with a monitor in the cab.

Any truck in disrepair should not be allowed on the job site. The following is a brief list of some items that should be checked:

1. Headlights, tail lights, brake lights, directional signals, speedometers, fuel gauges, horns, and windshield wipers, as well as cooling, heating, defrosting and defogging systems, rearview mirrors, and so on.
2. Nonslip access surfaces.
3. All glass in windshields and windows, which should be safety glass and promptly replaced when broken.
4. Tow bars, for adequacy.
5. Exhaust systems, for leaking fumes.
6. Fenders and/or mud flaps. Rear dual wheels have a splitter bar between the tires to prevent rocks from being caught and later ejected as speed is increased.
7. Emergency equipment: three emergency flares, reflective markers, or other equivalent portable warning device.

## Pickups

All pickups should comply with the above.

## Dump Trucks

All dump trucks should comply with the above. They should have a sturdy bumper bed body to protect the cab and the operator from falling rocks and other materials during the loading process, and the dumper bed should be strong enough to prevent the cab from being crushed in the event of rollover, as required by SAE J1040.[1] On many dump trucks the dump bed body extends over the truck cab, but this design does not give sufficient protection if the truck happens to overturn with the bed up and rolls down the spoil pile, in which case the cab is very vulnerable to crushing. Because a truck of this type was equipped with a separate ROPS on the cab when it overturned and made three complete rolls to end up on its wheels in the Stanaslaus River on the Corps of Engineers' New Melones Dam project in its Sacramento District in about 1979, the driver was able to walk away uninjured. All the truck needed was a change of oil, and both driver and truck were back on the job the same day.

All dump trucks should be equipped with a safety strut to prevent accidental lowering of a raised dump bed while service or inspection is being done.

All hoist levers for the dump bed should be designed to prevent inadvertent raising or lowering of the bed.

Trip handles for tailgates should be arranged so that the operator is in the clear when the gate is opened.

Because raised dump beds often have collided with overhead objects, it is wise to have an audible alarm installed so the driver will be warned that the bed is still up.

Dump trucks used on steep roads should be equipped with maxi-brakes that will automatically stop the vehicle if the air brake system should fail.

Fifth-wheel dump trucks are very vulnerable to side upset if the rear wheels under the dump bed are on a side slope, a rut, or soft ground not immediately visible to the truck driver. A simple slope indicator in the cab would give the driver the critical information needed to avoid an upset. These dump beds can be raised as high as thirty feet in the air, where they can come into contact with powerlines that the driver may not be able to see from the cab. To avoid this hazard, it is suggested that a proximity alarm be installed so that the driver will be made aware of the possibility of powerline contact.

### Large Heavy Haulers

This category includes flatbed, tank, stake (for pipes, poles, etc.), and tractor/fifth-wheel semitrailer trucks, as well as log trucks whose rear wheels are attached by a reach, cargo vans, and so on.

All large heavy haulers should be checked for the general items listed above.

For fifth-wheel and log trucks, a locking device or a double safety system should be provided to prevent inadvertent separation.

When operated on public highways, these trucks must comply with pertinent state laws and interstate commerce requirements.

All vehicles, except trailers or semitrailers having a gross weight of 5,000 pounds or less, should be equipped with service brakes and manually operated parking brakes. Service and parking brakes should be adequate to control movement, stopping, or holding of the vehicle under all conditions of service. Service brakes on trailers and semitrailers should be controlled from the prime mover's driver's seat.

Braking systems should be synchronized on all wheels or should be of the antilocking type. Braking should be on the rear wheels first to prevent jackknifing.

Stake trucks for transporting pipes, poles, or logs should have the stake release on the side opposite the unloading side to prevent injury due to rolling pipes, poles, or logs.

*Trailers*

This category includes two-wheeled trailers for carrying two wheels of another vehicle, small cargo trailers, small tilt-up trailers for loading and offloading, and large trailers connected by either a fifth-wheel or a hitch for lowboys, cargo, equipment, and other bulk hauling, and so on.

All trailers should be equipped with tie-down systems so that loads cannot shift during travel.

All trailers connected by a hitch need two safety chains connecting the truck and the trailer to prevent separation during travel.

Trailers used to carry wheeled vehicles should have curbs on the inside of the trackways to prevent side movement or sliding during loading or offloading. Cranes, excavators, and other heavy mobile equipment can slide sideways and overturn if a trailer happens to sink sideways into soft ground. (See Figure 14-1.) A four-inch curb on the inside of the trackways would prevent such sliding and would not impede side-loading or self-loading. In one case, a lowboy carrying a large track-mounted excavator was stuck in mud and tilted toward a pond. Apparently, in order to free the lowboy, the excavator operator tried to walk the excavator off the end of the lowboy. As he was doing this, the excavator slid sideways and off the side of the lowboy into the water, and the operator was drowned when he became trapped in the cab under water.[2]

**Figure 14-1.**   Excavator overturned.

To make tilt-up trailers more user-friendly, they should be equipped with a winch to assist in loading and offloading.

End ramps for driving heavy wheeled equipment on and off trailers should be well secured so that they will not fall off during travel and endanger other vehicles on the road.

A trailer with a drawbar should have a supporting third leg so that it can be safely lowered or raised when being connected or disconnected. Many workers have been injured by erratic vertical movement of the drawbar while performing this task.

## Not Elsewhere Classified

Three- and four-wheeled all-terrain vehicles (ATVs), currently popular with children and adults and used on farms, ranches, and construction, are exceedingly unstable. After over 600 deaths had been recorded, the American Academy of Pediatrics began discouraging the manufacturer of three-wheeled ATVs. Both three-wheelers and four-wheelers can easily do a "wheely," raising up the front end of the vehicle, which can result in loss of control and upset. Most of the associated deaths and injuries have occurred when the vehicles have rolled over, or the drivers have hit fence wires, guylines for power poles, low branches, or other things not readily visible to the operators. A four-post protective canopy frame system with seat belts would provide protection against ejection and entrapment. Some construction managers have restricted ATV use to specific areas and issued in-house licenses to a select few employees who have had training in operating ATVs.

## Forklifts

A conventional forklift is a handy workhorse on a construction project for moving materials short distances and for stacking materials for storage. Forklift masts sometimes are used as accessories on farm-type wheel tractors, skid-steer loaders, wheeled loaders, and many other types of vehicles for both lifting and transporting. All forklifts that can raise the load above the operator or are used where objects might fall on the operator need overhead falling object protective systems (FOPS). The protective canopy should be designed to resist the anticipated force of an upset because forklifts are often inadvertently driven off loading docks. Forklifts also are very vulnerable to side upset when making sharp turns, particularly when they are not carrying a load or the load is in a raised position, which can cause the operator to be ejected and crushed by the FOPS.

Most forklift manufacturers have developed and are installing on newer models operator restraint systems, which include seat belts and seat configurations that prevent sliding to the side, to confine the operator within the FOPS in the event of an upset. Specific warnings now are posted in forklifts instructing operators to use seat belts at all times and brace themselves if the forklift starts to be upset, so they will stay within the protective frame of the FOPS and not be ejected. Many operator training courses are available, which can be used as the basis for an in-house licensing program.

## Conveyors

Conveyor systems, used in rock crushing, gravel plants, asphalt plants, pumpcrete systems, and so forth, often are an important part of a construction project. The usual types of conveyors are the belt, roller, auger (screw), bucket, and trolley. All must be closely examined at the time of assembly for compliance with the National Safety Council's *Accident Prevention Manual for Industrial Operations* and *Data Sheets*[3] and ANSI B20.1, "Safety Standards for Conveyors and Related Equipment,"[4] which go into greater detail than OSHA requirements. The absence of a lockout system to deactivate a conveyor during maintenance or repairs has caused many serious injuries. The conveyor must be physically locked out by the individual performing the maintenance or repairs, who must have the only key to the lock. Another necessary feature is an emergency stop cord running the full length of the conveyor so that it can be immediately stopped by anyone who sees a problem. This device must be designed so that the conveyor cannot be restarted by someone else until certain procedures have been taken to ensure that whoever pulled the cord is not still on the conveyor. Provision should be made for safe access and egress to prevent falls. Guarding of accessible moving parts also is required. The auger-type conveyor is by far the most dangerous and should have an interlock to stop it whenever its cover or grate is opened, whether for service or otherwise. Many workers have been killed or have lost limbs when the screw of the auger pulled them in like a meat grinder.

## Pipelines

Water, compressed air, and fuel often are piped to various locations on a project. Pipelines should be located so that they are protected from vehicular traffic and will not create hazardous circumstances if a pipe should leak, burst, or otherwise fail. A professional engineer familiar with pipeline design requirements should certify the adequacy and the safety of all pipelines.

## Railways

Installation of railroad track should meet standards adopted by the American Railway Engineering Association (AREA) and the Federal Railroad Administration (FRA). An engineer familiar with rail installation should certify its adequacy for rolling stock to be used on it.

## Roadways

Layout of haul roads and other roadways should be consistent with ANSI D6.1, "Manual on Uniform Traffic Control Devices for Streets and Highways,"[5] particularly when the public has access to the roads, or a public road must make a detour through a road construction project. Haul roads on hillsides should be designed with emergency runaways and berms to prevent the loss of haul trucks. Do not mix off-highway hauling equipment with on-highway vehicles. A roadway should not cross or be utilized as an airstrip unless special provision is made consistent with requirements of the Federal Aviation Administration (FAA).

## Aircraft

FAA requirements apply to the use of all aircraft, airport facilities, and airdrops of materials.

## EARTHMOVERS AND EXCAVATORS

First on the agenda on most construction projects is to change the existing landscape. Land is cleared and reshaped by excavating and moving earth, removing or quarrying rocks, and sometimes digging tunnels.

Equipment used to do this should conform to the recommended practices of the SAE handbook for *On-Highway Vehicles and Off-Highway Machinery.*

Excavators, bulldozers, front-end loaders, graders, farm-type tractors, skid-steer bobcats, road rollers, compactors, street sweepers, scrapers, and all other mobile equipment require rollover protective systems (ROPS). (See Figure 14-2.) The ROPS include the structure, a seat belt, and warnings and must conform to performance criteria as set forth in SAE Standard J1040. For additional information about ROPS see Chapter 1 in *Automotive Engineering and Litigation,* Volume 2.[6] A scraper with a crowd-type loading mechanism should have a substantial wire grating installed behind its ROPS frame to protect the operator from rocks and heavy clods of dirt that may be thrown from the crowd into the open cab.

All mobile equipment should be equipped with backup alarms (reverse signal alarms).

Large excavators that will be working adjacent to water should have an alternate escape route from the cab so that the operator has a way out if overturned into the water (as in Figure 14-1). SAE J185, "Access Systems for Off-Road Machines," describes alternate systems to use when the primary access is not usable.

Drilling and tunneling equipment will be discussed in Chapter 23.

**Figure 14-2.** ROPS.

## LIFTING EQUIPMENT

Cranes, material and man hoists, and aerial lifts all fall within the regulatory requirements of OSHA and ANSI and the recommendations of SAE. Some people have the erroneous idea that they need only comply with the mandatory requirements of OSHA to achieve a safe workplace. This mind-set is a "recipe for disaster." Many effective hazard control devices have been developed to meet specific needs but are not included in OSHA, even though they have proved their life-saving worth and cost-effectiveness on many well-managed projects. The use of a safety device should be based upon its ability to save lives, not upon whether it is endorsed by OSHA or ANSI, as these bodies rely upon the long and tedious process of consensus rule-making, and the graveyard is usually full by the time something is written.

### Cranes

Specific information on crane hazards can be found in my book *Crane Hazards and Their Prevention,*[7] which discusses some fifty crane hazards and contains a worksheet for making a mobile crane hazard analysis, and thus is a useful guide for inspecting lifting equipment. Cranes brought onto the job site need to be examined to see if the safeguards discussed in the following paragraphs have been installed.

#### Proximity Alarms and Insulated Links

OSHA requirement §1926.550(a)(15)(v) refers to the much-debated efficacy of insulated links and proximity alarms. Tests have consistently demonstrated that any leakage of amperage on insulated links deliberately contaminated for the test is approximately the same as that acceptable for triggering ground-fault interrupters (GFI) and well below the paralysis threshold established by the International Electrotechnical Commission (IEC). An insulated link is capable of protecting a rigger guiding a load if the boom or hoist line should come into contact with a powerline. In Australia it is required that taglines be made of electrically nonconductive material.[8]

Proximity alarms have served as effective early warning devices in detecting powerlines, particularly on cranes used in pick-and-carry operations, as an operator of a crane with the boom raised may be momentarily unaware of powerlines in the path of travel because the operator's attention is focused on the load, not the boom tip.

No known personal injury litigation has arisen because of a failure of an insulated link or proximity alarm when used in accordance with the manufacturer's instructions.

In the section of this chapter on "Power Sources," additional hazard prevention measures for the avoidance of powerline contact are discussed.

### Load-Moment Indicators

This hazard is best controlled by installation of load-moment indicators (LMIs) or other load-measuring devices. (See Figure 14-3.) The margin of safety between the actual tipping load and the rated capacity often varies from 15 to 25 percent, based on the type of crane.

On some cranes, the location of the boom in relation to where the carrier is mounted determines the margin of safety between the tipping load and the rated capacity. For a crane mounted on a flatbed truck with the hydraulic boom located directly behind the truck cab, the tipping load over the rear of the flatbed or over the cab often can be as much as twice the rated capacity, but the crane can easily be upset when the load is slued (rotated) to either side. The weight

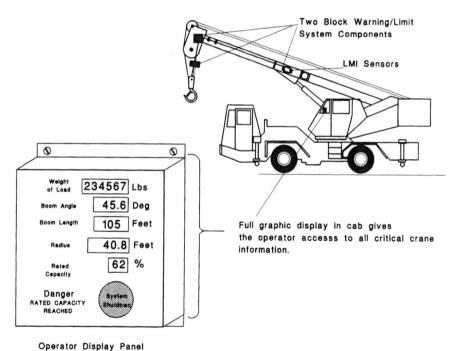

Full graphic display in cab gives the operator accesss to all critical crane information.

Operator Display Panel

**Figure 14-3.**   Loan-moment device (LMI).

of a boom without a load can create an imbalance and cause some high-reach hydraulic cranes to upset when the boom is positioned at a low angle. This has occurred even with outriggers extended.

A number of variables effect lifting capacity:

1. The ability to extend a hydraulic boom while raising or lowering it increases the radius very swiftly and reduces lifting capacity quickly.
2. Whether the operator chooses to extend or to retract outriggers affects the crane's stability.
3. The crane's tipping capacity can vary when the boom is positioned at the various points of the compass or clock in relation to its particular carrier frame.
4. In many instances, the operator may not know the actual weight of the load. Often the operator relies upon his/her perception, instinct, or experience to determine whether the load is too heavy, and may not respond fast enough when the crane begins to feel light.

All these variables create error-provocative circumstances for operators, and can lead to their inadvertently exceeding the rated capacity, or to tipping of the load, upset, or structural failure—all of which can be avoided by planning, training, and using an LMI.

Improved crane rating charts have helped operators to understand the lifting capacities of cranes. The chart in the crane should be consistent with SAE J1257, "Rating Chart for Cantilevered Boom Cranes."

### Anti-Two-Blocking Devices

Latticework and hydraulic-boom cranes are both prone to two-blocking and should be equipped with functioning anti-two-blocking devices. Two-blocking has caused hundreds of deaths and crippling injuries.

In walking a crawler crane with a long latticework boom without a load, a great deal of whip is created, and the headache ball and empty chokers can drift up to the boom tip. Ordinarily, while busy watching the pathway of travel to avoid any rough ground that can violently jerk the boom, the operator does not watch the boom tip. When a hoist line two-blocks, it picks up the weight of the boom and allows the pendant guys to slack. Then, if the crane crawler breaks over a rock or a bump, the flypole action of the long boom is sufficient to break the hoist line. If the hoist line on the jib of a latticework crane two-blocks, it can pull the jib up and over backwards, and pull it from the tip of the main boom, causing the whole jib to fall.

The power of the hydraulic rams that extend hydraulic booms often is sufficient to break the hoist line if the line two-blocks.

Some latticework and hydraulic-boom cranes will two-block when the hook is near the tip and the boom is lowered.

Anti-two-blocking devices are now generally standard on all new cranes. (See Figure 14-4.) For older cranes, they are available from most crane manufacturers and after-market suppliers.

*Safety Latches*

Safety latches are necessary on every crane hook used on a construction site. A safe lifting hook is a critical component in lifting a load, as a hook that has no latch to secure the load straps or chains within the throat of the hook is a killer. Many hooks have thin sheet metal latches that are easily damaged or bent and are totally worthless for securing the straps safely within the hook.

Before a crane or other lifting equipment using hooks is allowed to operate on the job site, make sure that the hook has a double-locking safety latch. A

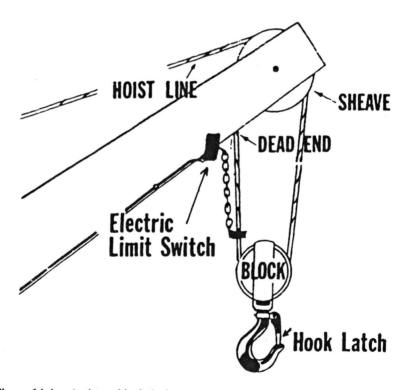

**Figure 14-4.**   Anti-two-block device.

vertical swivel also should be used so that the hook will always remain vertical and place no side pressure on the safety latch. Over 100 positive-type safety latches designed to prevent straps from slipping out of the throat of the hook have been patented, many of which are available in the market. (See Figure 14-5.)

Many excavators with hydraulic booms use a backhoe bucket that has a lifting hook for placing pipe into the trench being dug. When the bucket is turned down underneath the boom, the throat of the hook can turn downward, allowing the straps to slip out of the hook. To avoid this hazard, a ring instead of a hook should be used, with a shackle to connect the lifting straps.

### Damaged Cable

All crane cables should be reeled out so that they can be thoroughly examined, and damaged cable can be replaced. Cable damage usually can be traced to worn and damaged sheaves that are abusing the wire rope as it passes over them. Worn and damaged sheaves also can cause cables to snag

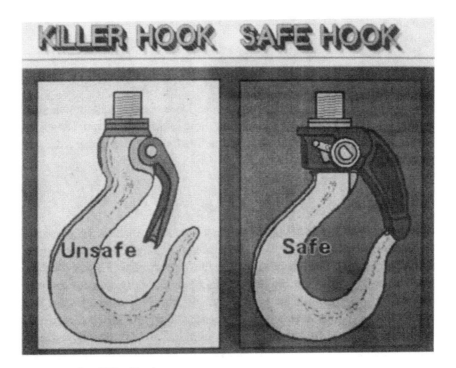

**Figure 14-5.**   Killer Hooks.

and momentarily hang up, severely damaging the cable. Inspect cable daily so that wire rope can be taken out of service when six randomly distributed, broken wires occur in one lay, or three broken wires in one strand in one lay; inspect the sheave diameter and the groove to determine if they match the cable size specified by the crane manufacturer; and be sure that cables are properly lubricated to reduce wear.

### Boom Stops

Cranes with latticework booms should be equipped with two types of boom stops. The first type is an electrical interlock sensor-type system that intercedes and stops boom movement beyond a vertical angle of approximately 86 degrees and prevents the boom hoist mechanism from raising the boom any farther. Be sure to see that this type of system is functioning. The second type is a mechanical bumper consisting of two energy-absorbing cylinders that attach the lower section of the boom to the frame of the cab structure. Some of these cylinders are spring-loaded devices that offer nominal protection. The most effective barrier-type boom stops have energy-absorbing hydraulic cylinders that stabilize the boom. Hydraulic boom stops also overcome most of the force created by sudden dropping of a load and can aid in overcoming some of the force of unexpected wind gusts.

### Spirit Level

The cab of a crane should be equipped with a spirit level so that the operator can level the crane horizontally in all four directions. With no spirit level, an operator may be unaware that the crane is tilted, which can cause side pull and reduce lifting capacity.

### Crane Inspection

OSHA requires in §1926.550(a)(6) an annual crane inspection by a competent person recognized by the U.S. Department of Labor. It is necessary to examine very closely the checklist used in the inspection to find out what the inspection covered and the findings of the load test, to make sure that the crane will be suitable for its intended use on the project. Private agencies that make these inspections have no enforcement authority to ensure that necessary repairs are made before the crane is used. Check very closely to see if the deficiencies listed have been remedied. Crane inspection agencies are generally very cooperative and will make reexaminations to verify compliance with previous inspections and with special requirements deemed appropriate by the safety plan.

## Material and Man Hoists

Material and man hoists must meet OSHA and ANSI requirements. OSHA 1926.552 provides basic guidance for selecting safe lifting equipment. Elevator standards provide additional insight for achieving a high level of safety and reliability when hoists are used. Because it is human nature to ride rather than to climb stairs, be sure that any material hoist used on the job site is designed so that it also will carry people, to avoid a policing problem.

The erection of hoists should be under the supervision of a qualified registered engineer competent in this field.

## Aerial Lifts

Because of the great number of injuries and deaths due to contact of powerlines by aerial lifts or work platforms, planners should consider using lifts with nonconductive masts, booms, or ladders. A number of workers have been crushed against overhead objects when their work platforms have been unexpectedly raised because an unprotected control lever has been activated by another object.

Aerial lifts with outriggers should have an interlock system to prevent raising of the boom when outriggers are not extended and in place, to prevent unintentional upset.

Some self-propelled aerial lifts rely upon counterweights rather than outriggers. They must have a functioning interlock to prevent use of the boom if the lift is not on level ground, to prevent upset.

## SCAFFOLDING AND TEMPORARY SUPPORT STRUCTURES

Approximately one-third of all construction injuries and deaths involve falls from elevations; so the preventon of falls from elevated areas should be of primary concern in all construction planning. Temporary or permanent installation of scaffolding for access ways, ramps, staging, and support structures is vital to a safe construction site (OSHA §1926.451 and §1926.500). A registered professional engineer should examine, sign, and seal design drawings of such structures to certify that they are adequately designed for their intended use. Portable prefabricated steel-pipe scaffolding needs to be examined closely before use. Discard those sections that are broken, bent, or damaged. Particular care should be taken in examining the leveling components and castors to see that they function properly. (See Chapter 12 for additional guidelines.)

If proposed construction methods make it physically impossible to install scaffolding to give workers a safe work platform, then a secondary plan should be developed to provide fall protection. For additional information, see Ellis's *Introduction to Fall Protection.*[9]

If scaffolding is fabricated on the job to serve as a temporary support structure for concrete forms or steel structural members that need to be assembled in place, be sure the design is approved and certified by a registered professional engineer who is qualified to make such assessments.

## POWER SOURCES

Construction planning must include consideration of the power sources needed on the project. Almost all operations rely upon electric power and its distribution system. Rock drilling and some air-powered tools rely heavily upon compressed air. Fueling stations for gasoline and diesel often are included on very large projects. All of these systems have inherent hazards that must be addressed in construction planning.

### Electrical

Because powerlines do not mix with boomed and high-clearance equipment, the location of existing power distribution systems and proposed permanent or temporary powerlines on or adjacent to the job site must be reviewed before any work is commenced.

Figures 14-6 and 14-7 show the safe positioning of cranes in relation to powerlines so that it is impossible for the boom to come into contact with a powerline. Removal or burial of powerlines in locations where cranes will be used is a must in construction planning. One of the first priorities of the construction manager is to notify the electric utility so that it can participate, if possible, in project safety planning (Chapter 6), the prebid safety conference

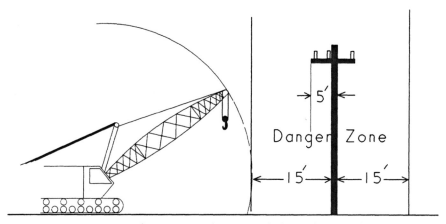

Crane boom is lowered level to the ground
and boom tip cannot intrude into the danger
zone extending 15 feet on each side of the
power line.

Safe Crane Location (Side View)

**Figure 14-6.**   Cranes and powerlines, side view.

(Chapter 10), and the pre-notice-to-proceed conference (Chapter 13), to create a clear undestanding and firm agreement on the separation of cranes and other equipment from powerlines. The *National Electrical Safety Code* does not specify any safe vertical clearance for powerlines in construction activities, as it does for rural and urban roads where vehicle clearances are known. Prudent interpretation of this code would dictate that powerlines should not be in or adjacent to the construction areas.

Over the years thousands of crippling injuries and deaths have resulted from contact between cranes and powerlines. Human factors specialists have found that it is simply beyond the range of normal human performance to be able, visually, to judge clearances between a crane boom or hoist line and powerlines with any accuracy because the sky's bright background fails to provide a visual parallax. They have also found that workers cannot observe more than one visual target at a time or overcome the camouflaging characteristics of trees, buildings, and other objects.[10,11] The visual estimates of clearance made by crane operators, riggers, and signalers are not reliable as they need visual clues placed on the ground to identify the danger zones created by powerlines. (See Figure 14-8.)

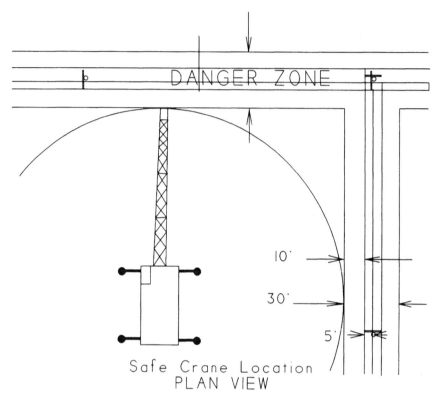

**Figure 14-7.**  Mapping.

It is like signing a death warrant to store material under powerlines, as someone may choose to lift the material with a crane. Construction planning must designate a thirty-foot-wide danger zone under all powerlines to declare this area off limits for material storage, as shown in Figure 14-6. This thirty-foot area should be enclosed by fencing fifteen feet out from the centerline measured from the power poles.

All electrical facilities on a project, whether temporary or permanent, must be in compliance with the *National Electrical Safety Code* and the *National Electrical Code* and must be designed and installed under the supervision of a registered professional electrical engineer.

## Compressed Air

Compressors, air receivers, piping, and hoses used to carry compressed air for pneumatic tools must meet specific requirements set forth by the Ameri-

# DANGER

### YOU WILL BE
# ELECTROCUTED
### FROM BOOM—POWERLINE CONTACT

## DANGER ZONE
### UNSAFE FOR CRANE OPERATIONS
**DO NOT lift or make boom movements inside the barricaded area.**

Map and Barricade the 30 foot wide Danger Zone
(15 feet on each side of the powerline poles).

**ALWAYS notify the power company before you begin crane
operations near powerlines.**

Power Company                    Telephone

**Figure 14.8**

can Society of Mechanical Engineers' (ASME) *Code and Manual of the National Board of Boiler and Pressure Vessel Inspector.* All proposed compressed air systems, as well as all pneumatic tools to be used on the project, should be viewed by a licensed registered professional mechanical engineer to assure their safety and compliance with appropriate standards, requirements, or regulations.

All rock-drilling equipment brought onto the project should be examined closely to ascertain that it is safe for handling compressed air. All hoses and temporary piping should have safety chains to prevent them from flailing about if they become disconnected under pressure. It is a dangerous practice to pressurize a water tank on a trailer to force water through hoses to suppress dust created by rock drilling. Because these water tanks usually are not designed or intended to be used as pressure vessels, they have been known to explode, causing death and injury.

If the type of drilling equipment needed has a high mast tall enough to be raised into powerlines, the powerlines must be removed from the drilling site before drilling is commenced.

## *Fuel*

Some projects have their own gasoline, diesel, or liquefied petroleum fueling facilities. These must be installed in strict conformance with the *National Fire Codes*, Environmental Protecton Agency rules, and local requirements and under the supervision of a registered professional engineer competent in this field.

## MATERIAL PROCESSING

When concrete and asphalt batch-plants, gravel and aggregate plants, roto-mills for reprocessing pavement, paving machines, and many other types of material-processing machinery or systems are used on construction projects, a qualified safety engineer should examine the equipment when it is completely assembled and before it is used to assure its safety and its compliance with recommendations contained in the manufacturer's operating manual. Caution is required because the equipment as manufactured may not have proper guarding for accessible moving parts. This equipment also should have emergency stops and lock-out systems. For complex plants, a system safety hazard analysis may be desirable, as they may have error-provoking controls or other problematic features.

## MARINE EQUIPMENT

Some construction projects involve work around, on, or under water. The U.S. Coast Guard requirements apply to all floating plants, including boats, barges, dredges, and diving or other water-related equipment. The U.S. Army Corps of Engineers' *Safety and Health Requirements Manual* EM 385-1-1, in Section 19 covering floating-plant and marine activities, is another good source of information.

## HAND TOOLS

Many injuries result from unsafe or improper use of hand tools, particularly those that are powered. All hand tools must be inspected before use for their suitability for the proposed work and to eliminate hazardous tools from the job site.

Extension cords for electrical tools must have ground-fault interrupters. The integrity of insulation on double-insulated electric tools must be tested before their use is permitted, and a plan must be implemented for monitoring them during the life of the project.

Pneumatic saws and grinders should have automatic overspeed controls to avoid runaway speeds that can cause carborundum saw discs and buffers to disintegrate, and guards must be in good condition to restrain flying particles.

All tools that produce dust need exhaust systems to collect the dust, to prevent contamination of the air in the work areas. Automatic emergency valves should be installed at all compressed air sources to shut off the air immediately if the hose becomes disconnected or is severed, to prevent wild thrashing of the hose.

## MISCELLANEOUS—NOT ELSEWHERE CLASSIFIED (NEC)

Safety belts, fall protection, goggles, respirators, and other personal protective equipment fall into this category, as do toxic fume detectors, emergency communications equipment, office furniture, and other items not otherwise classified that are used on a construction project; and all should be examined for adequacy and safety.

The principles outlined in this chapter can be applied to the examination of all equipment and tools brought onto the job site, not just the ones discussed. The safety engineer who conducts or oversees the examination of equipment and tools should research, review, and study as many references, texts, and standards as possible in preparing the outline of applicable requirements.

## REFERENCES

1. Society of Automotive Engineers, "Performance Criteria for Rollover Protective Structures (ROPS) for Construction, Earthmoving, Forestry and Mining Machines," J1040; and *On-Highway Vehicles and Off-Highway Machinery, Handbook,* Vol. 4.

2. MacCollum, David V. *Crane Hazards and Their Prevention,* American Society of Safety Engineers, Des Plaines, IL, 1993.

3. National Safety Council, "Engineering and technology," *Accident Prevention Manual for Industrial Operations,* Chap. 4; and *Data Sheets* 569 and 570.

4. American National Standards Institute, "Safety Standards for Conveyors and Related Equipment," B20.1, U.S. Government Printing Office, Washington, DC.

5. American National Standards Institute, *Manual on Uniform Traffic Control Devices for Streets and Highways,* D6.1, U.S. Government Printing Office, Washington, DC.

6. MacCollum, David V., "Rollover Protective Systems (ROPS)," *Automotive Engineering and Litigation,* Vol. 2, Chap. 1, Peters and Peters, Editors, Wiley Law Publishing, Colorado Springs, CO, 1988.

7. MacCollum, David V., *Crane Hazards and Their Prevention,* American Society of Safety Engineers, Des Plaines, IL, 1993.

8. Queensland, Australia, Division of Workplace Health and Safety, *Civil Construction Safety Manual,* paragraph 7.7(9), 1991.

9. Ellis, J. Nigel, *Introduction to Fall Protection,* American Society of Safety Engineers, Des Plaines, IL, 1994.

10. Middendorf, Lorna, "Judging clearance distances near overhead powerlines," Human Factors Conference, Detroit, MI, 1978.

11. Cunitz, Robert J. and Middendorf, Lorna, "Problems in the perception of overhead powerlines," 6th International System Safety Conference, Houston, TX, 1983.

Chapter *15*

# *Fall Prevention and Protection*

When one falls, it is not one's foot that is to blame.

(Chinese proverb)

"Fall prevention" and "fall protection" are not the same, even though the terms commonly are used synonymously. Fall prevention means the elimination of hazards that cause falls by building safe platforms enclosed by proper railings so that there is no cause to fall from an elevated area. Part of fall prevention is to provide a nonslip and level walking surface that has no holes or protrusions. Fall protection is necessary when it is impossible to provide safe platforms and railings. Fall protection includes safety belts, harnesses and lines, shock-absorbing lanyards, and safety nets. Construction safety planning should never substitute fall protection for fall prevention when fall prevention is a feasible option. Falls from an elevation account for about 22 percent of all construction injuries and deaths. Three-fourths of the injuries and deaths sustained by ironworkers can be attributed to falls from an elevated area.

## FALL PREVENTION

Falls are classified into two groups: falls from an elevation and falls on the same level. Because falls from an elevation are the more serious of the two

*95*

types, construction safety planning should require scaffolding, work platforms, decking, and other staging to be enclosed fully by proper railings, thus limiting falls to the same level and reducing the severity of injury. This is not to say that falls on the same level cannot be extremely dangerous, as a worker might fall and strike a sharp edge or be stabbed by rebar or some other pointed object. When you see an orange mushroom patch growing on forms for ground-floor footings, you know the supervisor has recognized that a worker who makes a simple misstep could be impaled on the eighteen-inch-high bare pieces of upright rebar, even though well-defined walkways are in place. These reusable, plastic, mushroom-shaped rebar covers are an excellent investment in fall protection, as they can not only save the life of someone who falls on the same level but also save one who falls from an elevation. Of course, individuals working at higher levels should be protected by a safe work platform or some other form of fall protection.

There are many, many references for the design of safe scaffolds, work platforms, and other staging to prevent falls from an elevation. A professional engineer familiar with structures of this type should approve their design. Safe staging is very cost-effective, as it allows a worker to use both hands instead of having to hold onto something with one hand as a brace, which is what happens one uses a safety belt or harness hooked to a safety line or a fixed structure. This also eliminates the supervisor's having constantly to remind workers to be properly secured. Cost studies also show that work progresses faster from safe staging, which is a profit-producer as the initial cost of the staging usually is quickly recouped.[1]

Accessways should be free from obstruction and protected from the elements and should have nonslip surfaces to reduce the risk of falls. Also the following checklist should be helpful in safety planning:

1. Safe access to work areas includes provision for:
   a. A stairway, ladder, ramp, or personnel hoist where there is a horizontal or a vertical break of 19 inches or more in a route of access.
   b. Nonmetal accessways where electrical work is to be done or where the accessway might contact electrical conductors.
   c. Keeping accessways clear to permit free passage of workers. If the accessway must be temporarily obstructed because of the nature of the work, a second means of access should be provided, particularly if it involves access from one work level to another.
2. The project safety engineer should make an initial hazard analysis of access needs. This hazard analysis should be approved by the construction manager and include:

    a. The design, construction, and maintenance of the means of access.

    b. Erection and dismantling procedures, including provisions for providing fall protection during the erection or dismantling when erection or dismantling involves work at heights six feet or more above the ground.

    c. Requirements for job-made means of access. These need to be designed to support, without failure, at least four times the maximum intended load.

    d. The width of accessways. Widths are dependent upon the purpose for which they are built and should be sufficient to provide safe passage for materials and workers. Except for ladders, the width should never be less than 18 inches.

3. When wooden scaffolds are to be used, the following guidelines apply:

    a. Solid sawn planks should be of a scaffold-plank grade and certified by, or bear the grade stamp of, a grading agency approved by the American Lumber Standards Committee.

    b. Planking should be secured to prevent loosening, tipping, or displacement, and supported or braced to prevent excessive spring or deflection; intermediate beams should be provided to prevent dislodgment of planks due to deflection.

    c. The spans listed below are permitted for individual wood planks of scaffold-plank grade:

        (1) Douglas Fir or Southern Pine: 2" × 10" nominal or $1\frac{1}{2}" \times 9\frac{1}{4}"$ actual. Maximum permissible span: one worker, ten feet; two workers, eight feet; and three workers, five feet.

        (2) Spruce, rough-sawn, 2" × 9" nominal or $1\frac{7}{8} \times 9\frac{7}{8}$ actual. Maximum permissible span: one worker, eight feet; two workers, seven feet; and three workers, five feet.

    d. Planking should be laid with edges close together across the entire access surface so that there are no spaces through which personnel, equipment, or material could fall.

    e. When planking is lapped, each plank should lap its support at least twelve inches and be over a cross joist.

    f. All load-carrying lumber used for accessways, except planking, should be a minimum of Number 1 Southern Pine grade, Number 1 Douglas Fir, or the equivalent:

        (1) All dimensions are nominal sizes (except where rought sizes are noted) as provided by the "NBS Voluntary Product Standard PS 20-70, American Softwood Lumber Standard," published by the National Bureau of Standards of the U.S. Department of Com-

merce: where rough sizes are noted, only rough or undressed lumber of the size specified will satisfy minimum requirements.

(2) Lumber should be reasonably straight-grained and free of shakes, checks, splits, cross grains, unsound knots or knots in groups, decay and growth characteristics, or any other condition that will decrease the strength of the material.

g. Supporting members and foundations should be of sufficient size and strength to distribute loading safely:

(1) Supporting members should be placed on a firm, smooth foundation that will prevent lateral displacement.

(2) Unstable objects such as barrels, boxes, loose bricks, or concrete blocks should not be used as supports.

h. Poles, legs, uprights, or other vertical members should be plumb and securely braced to prevent swaying or displacement.

i. The design and construction or the selection of planking and platform should be based upon either the number of persons for which they are rated or the uniform load distribution to which they will be subjected, whichever is the more restrictive.

4. Accessways should have overhead protection equal to two-inch solid planking whenever work is performed overhead and workers are exposed to falling objects.

5. Nails should be driven full length. Double-headed nails should not be used on decks, guardrails, or handrails.

6. Planning should designate who will daily inspect accessways and who will maintain them so that:

a. Accessways are kept free of ice, snow, grease, mud, debris, or any other material or equipment that could obstruct passage, cause a tripping hazard, or render them unsafe in any other way.

b. They are not slippery. Abrasive material should be used for safe footing.

c. All obstructions or projections are removed or conspicuously marked. Obstructions or projections that are sharp, are pointed, or may cause lacerations, contusions, or abrasions should be covered with protective material.

d. When they become damaged or weakened, they are not used until repaired or replaced.

7. All accessways, ramps, and platforms should have guardrails that meet the following specifications:

a. Toprails, midrails, and posts shall have a vertical height of forty-two inches from the upper surface of the toprail to the floor level. Standard guardrail systems should have towboards on all open sides and ends to prevent objects from falling off.

b. Toprails and midrails should be smooth-surfaced.

c. Midrails should be halfway between the toprail and the floor.

d. Ends of toprails and midrails should not overhang the terminal posts.

e. Synthetic or natural fiber ropes should not be used as toprails or midrails. Wire rope may be used if tension is maintained to provide not more than four inches of deflection in any directon from the centerline, under a 200-pound weight and if support posts are located nor more than eight feet apart.

f. Dimensions should be as follows:

(1) For wood railings, toprails should be of at least 2″ × 4″ stock, and midrails should be at least 1″ × 6″ stock.

(2) For pipe railings, toprails and midrails should be at least 1″ × 0.070″ wall steel tubing or 1.990″ × 0.058″ wall aluminum tubing.

(3) For structural steel railings, toprails and midrails should be of 1¼″ × 1¼ × ⅛″ angles or other metal shapes of equivalent strength, spaced not more than eight feet on center.

(4) For pipe railings, posts should be at least 1″ × 0.070″ wall steel tubing or 1.990″ × 0.058″ wall aluminum tubing, spaced not more than eight feet on center.

(5) For structural steel railings, posts should be of 1¼ ″ × 1¼″ × ⅛″ angles or other metal shapes of equivalent strength, spaced not more than eight feet on center.

g. Because sometimes suppliers of building materials remove handrails when raising loads of sheetrock, block, or other supplies to upper levels of a building being built or repaired, someone needs to be assigned to see that the handrails are replaced and *replaced properly*. One worker who leaned against an improperly replaced handrail fell because the supplier renailed the rail in the original holes, which did not hold.

8. The anchoring of posts and framing of members for all guardrails should be of such construction that the completed structure should withstand a load of at least 200 pounds applied in any direction at any point on the toprail without failure and with a minimum of deflection.

9. Toeboards should be constructed that are:

a. Made of 1″ × 4″ lumber or the equivalent.

b. Securely fastened in place and have not more than ¼″ clearance above floor level.

    c. Made of any substantial material, either solid or with openings not over one inch in the greatest dimension.

Where material is piled to such a height that a standard toeboard does not provide protection, paneling or screening from the floor to the toprail or the midrail should be provided.

10. Stairways should be used whenever possible to provide workers with safe access from one level to another, as they are a safer choice than either ramps or ladders. They should meet the following design criteria:

    a. Temporary stairways should have landings that extend not less than thirty inches in the direction of travel and are at least twenty-two inches in width for every twelve feet or less of vertical rise.

    b. Stairs should be installed between 30 and 50 degrees from the horizontal.

    c. Risers should be of uniform height and treads of uniform width.

    d. Metal pan landings and metal pan treads, when used, should be secured in place and filled with concrete, wood, or other material at least to the top of each pan.

    e. Wooden treads should be nailed in place.

    f. Every flight of stairs with four or more risers or rising more than thirty inches should have standard stair railings or handrails:

        (1) On stairways less than forty-four inches wide having both sides enclosed, at least one standard handrail should be installed, preferably on the right side descending.

        (2) On stairways less than forty-four inches wide having one side open, at least one standard stair railing should be installed on the open side.

        (3) On stairways less than forty-four inches wide having both sides open, one standard stair railing should be installed on each side.

        (4) On stairways more than forty-four inches wide, but less than eighty-eight inches wide, one standard handrail should be installed on each enclosed side and one standard stair railing on each open side.

        (5) On stairways eighty-eight or more inches wide, one standard handrail should be installed on each enclosed side, one standard stair railing on each exposed side, and a standard handrail in the middle of the stairway.

Portable stepladders should be limited to sixteen feet in height. Workers should not stand on the upper four steps, but should use them for bracing themselves while working. Stepladders should be used for limited tasks that require a minute or two to accomplish. Stepladders are not the safest equipment, as they are prone to be unstable because of uneven floors or debris, or are often in poor repair. The unsafe use of ladders has been a significant cause of injury. Ladders that are damaged or defective need to be removed from service immediately. Also, ladders should not be used as work platforms, but should be used only for temporary access from one level to the next. They should be secured on the top level to prevent lateral displacement, and should extend three feet above the next level.

## FALL PROTECTION

When fall protection is needed, an excellent reference is Ellis's book *Introduction to Fall Protection*,[2] which provides detailed information on the proper use of fall protection systems. The U.S. Department of Labor published in the Tuesday, August 9, 1994 *Federal Register,* 29CFR Parts 1910 and 1926, new OSHA standards for fall protection. This booklet of over 180 pages discusses in detail the supporting information upon which these new requirements were based, principally in Subpart M, Section 1926.500 and in the admendments in Part 1910.[3]

Figure 15-1 shows fall protection for concrete construction. The same concept can be used for steel erection.

Some general guidelines for the use of fall protection systems are as follows:

1.  Personal fall protection devices, independently attached or attended, should be used for performing various types of work from an elevated location, six feet or more above the ground, such as:
    a.  Work in hoppers, bins, silos, tanks, or other confined spaces.
    b.  Work on hazardous slopes, structural steel, or poles.
    c.  Erection or dismantling of safety nets.
    d.  Tying of reinforcing bars.
    e.  Work from boatswain's chairs, swinging scaffolds, or other unguarded locations.
    f.  Work on skips and platforms used in shafts by crews when the skip or the cage does not block the opening to within one foot of the sides of the shaft, unless cages are provided.
2.  Personal fall protection devices should be used as follows:
    a.  Body belts, body harnesses, or lanyards should be used for restraint

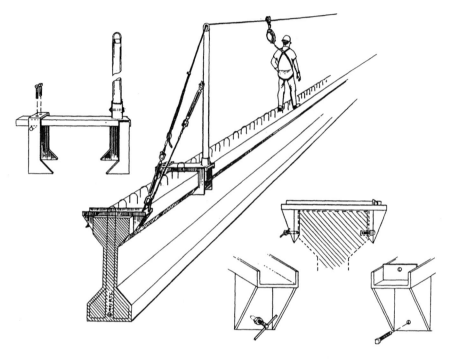

**Figure 15-1.** Fall protection, concrete beam. (Courtesy of the Research and Trading Corporation.)

and/or fall arrest where vertical free fall hazards exist and to reduce the probability of falls.

b. Lanyards should limit the fall to three feet; if the lanyard allows the worker to drop more than three feet, a shock-absorbing system (deceleration device) should be employed. A lanyard and a belt should be used only at the working level.

c. Where a possibility exists for a vertical free fall of more than three feet, chest harnesses, positioning and suspension belts, body belts, body harnesses, or lanyards should be used for suspension, support, and or positioning, and a shock-absorbing system (deceleration device) should be used.

3. Selection of personal fall protection equipment should be coordinated with the project safety engineer and should be based on: the type of work; the work environment; the weight, size, and shape of the user; the type and position of anchorage; and the length of the lanyard.

4. Personal fall protection equipment should be used for worker safeguarding. Any such equipment actually subjected to impact loading should be

immediately removed from service and should not be used again for worker safeguarding.

5. The anchor point for a lanyard or a shock-absorbing system (deceleration device) attached to a lifeline, or a lanyard or a shock-absorbing system (deceleration device) attached to a fixed anchorage point, should, if possible, be located above the wearer's belt or harness attachment.

6. A suitable anchorage point is one that is strong enough to support at least two times the potential impact load of a worker's fall, allows free movement of the attachment, is located as high as possible to prevent contact with an obstruction below should the worker fall, and is as vertical as possible in order to reduce swinging.

7. If a structural member with a sharp edge is used as an anchorage point for a rope, the rope should be effectively sleeved to prevent its being cut or abraded.

8. Body belts for fall arrest should be worn with the lanyard or the shock-absorbing system (deceleration device) attachment point(s) positioned in the center of the wearer's back at hip level. Body harnesses used for fall arrest should be worn with the attachment on the back of the wearer near the shoulders.

9. Safety nets should be used as follows:

   a. In unguarded workplaces over water, machinery, or dangerous operations, or more than twenty-five feet above the ground, the nets are used. Safety nets are not required for short-term repair work where other protection is provided.

   b. The use of personal fall protection devices (safety belt/harness, lanyard, and lifeline) is generally not an acceptable substitute for nets. However, for a specific operation, when approved in writing by the construction manager and delineated in the activity hazard analysis as discussed in Chapter 15, their use may be substituted.

   c. An operation requiring safety net protection should not be undertaken until a net is in place and has been tested.

   d. Safety nets should be installed as close under the work surface as practical but in no case more than twenty-five feet below such a work surface. Nets should be hung with sufficient clearance to prevent contact with the surfaces or structures below. Such clearance should be determined by impact load testing.

   e. Nets should extend at least fifteen feet beyond the edge of the work surfaces where workers are exposed.

   f. Forged steel safety hooks or shackles should fasten a safety net to its supports.

g. Connections between net panels should develop the full strength of the net.

10. Debris nets, when used with personnel safety nets, should be secured on top of the personnel safety net but should not compromise the design, construction, or performance of the personnel nets.

11. The project safety engineer should make sure that the debris net is of sufficient size and strength to catch expected debris, considering its weight, size, and length of fall.

12. Debris should be removed from safety and debris nets frequently. Nets should be protected from sparks and hot slag resulting from welding and cutting operations.

13. Safety and debris nets should be inspected by a competent person authorized by the construction manager in accordance with the manufacturer's recommendations. Fall protection should be provided those inspecting nets or removing debris. The frequency of inspections should be increased in proportion to the potential for damage to the nets.

## REFERENCES

1. MacCollum, David V., "How proper scaffolding cuts costs," *Western Construction*, Sept. 1957.

2. Ellis, J. Nigel, *Introduction to Fall Protection*, 2nd Edition, American Society of Safety Engineers, Des Plaines, IL, 1994.

3. "Safety Standards for Fall Protectio in the Construction Industry; Final Rule," 29 CFR Parts 1910 and 1926, *Federal Register*, Part III, Department of Labor, Occupational Safety and Health Administration, Tues., Aug. 9, 1994.

Chapter *16*

# Toxic or Hazardous Materials, Confined Spaces, and Respiratory Protection

Poison—Any agent which, introduced into an organism, may chemically produce an injurious or deadly effect.

*(Merriam Webster's Collegiate Dictionary, Fifth Edition)*

Today's world is filled with both toxic and hazardous materials that pose a serious threat to personnel safety and the global environment. Toxic or hazardous materials must be strictly controlled on any construction site. OSHA 1910.1200, "Hazard Communication," specifically requires that a "Material Safety Data Sheet," be prepared for hazardous chemicals. OSHA Form 200 has been developed for this purpose and outlines procedures for safe use and cleanup of spills and leaks. Consult the following references for assistance and further information:

1. The manufacturer.
2. *The Toxic Substance List* published by the U.S. Department of Health and Human Services.[1]
3. *Dangerous Properties of Industrial Materials* by N. I. Sax.[2]
4. Occupational Safety and Health act (OSHA), Part 1910.1000, Subpart Z, "Toxic and Hazardous Substances," through Part 1910.1500.[3]

## HAZARDOUS SUBSTANCES

Safety planning for the control of toxic or hazardous substances should include the following measures, taken from the U.S. Army Corps of Engineers' *Safety and Health Requirements Manual,* EM 385-1-1:[4]

1. Exposure to any chemical or physical agent via inhalation, ingestion, skin absorption, or physical contact in excess of the acceptable limits specified in the American Conference of Governmental Industrial Hygienists' (ACGIH) *Threshold Limit Values and Biological Exposure Indices* should be prohibited.
2. In the event of conflicts between ACGIH and other standards or regulations, the more stringent should prevail.
3. The construction manager should see that all contractors comply with all applicable standards and regulations to reduce contaminant concentration levels to as low as is reasonably achievable.
4. The project safety engineer should do the following:
   a. Evaluate all operations, materials, and equipment to determine the presence of hazardous environments or if hazardous or toxic agents might be released into the work environment.
   b. Evaluate operations, materials, and equipment involving potential exposure to hazardous sustances, agents, or environments to determine when a qualified industrial hygienist may be needed to formulate a hazard control program.
   c. Maintain a list of hazardous substances, agents, and environments, and recommend actions for their control.
5. A monitoring program should include:
   a. Use of approved and calibrated detection devices.
   b. Trained and competent monitors.
   c. A record of findings and actions taken.
6. An order of precedence will be established for control of toxic and hazardous substances as follows:
   a. Eliminate the substances when possible.
   b. Minimize exposure to toxic and hazardous substances by ventilation, wetting, and so on.
   c. Use respirators, gloves, or other special clothing.
7. When hazardous substances are procured, be sure that manufacturers provide Material Safety Data Sheets (MSDS) for them, make them available at the work site, and ensure that appropriate training is given for safe

handling, transporting, storage, and use. A person should be designated who will be responsible for compliance with the MSDS.

8. Facilities will be provided for quick drenching or flushing of the eyes and the body within the work area for immediate emergency use.

9. Disposal of surplus or excess materials and containers should be consistent with Environmental Protection Agency (EPA) requirements.

## RADIOACTIVE MATERIALS

Radioactive materials must be properly controlled when brought onto a site, and machines that produce radiation should be designed so that exposure of personnel is kept to a minimum. In developing a safety plan for the use of radioactive materials or machines that produce radiation, the following guidelines should be included:

1. Prior to initial storing, handling, using, transporting, and disposing of any radioactive materials or operation of a machine that produces radiation, the appropriate government authority should be notified of the proposed location, nature, and scope of the operation and provided a copy of any license required under Title 10, *Code of Federal Regulations* (CFR), Part 30. When radiation-producing machines or radioactive materials are used on a Department of Defense installation, a service permit is required. A lead time of forty-five days should be allowed for obtaining permits.

2. A qualified person should be designated as responsible for radiation safety for overseeing operations involving radiation hazards, conducting surveys and evaluations, and securing whatever specialized assistance is required to be in compliance with radiation protection standards. For materials used under Nuclear Regulatory Commission (NRC) license, only persons actually authorized by the NRC should oversee such work. Title 10, CFR, Part 20 governs the use of radioactive materials, permissible doses, levels of radiation, concentrations, precautionary procedures, and disposal.

3. Employees should not be exposed to a radiation dose in excess of that prescribed for the general public. Employees under eighteen years of age and pregnant women must not be allowed in areas where radioactive materials or radiation-producing machines are being used. Measurement of radiation exposure of body parts, including the head or the trunk, active blood-forming organs, the lens of the eye, and gonads, should be based upon the type of radiation and the energy produced, whether it is delivered simultaneously or successively, and the individual's age and accumulated occupational exposure. Records should be maintained for all persons

exposed to radiation where the dosage will exceed one-fourth the permissible limits and for each individual who enters a high radiation area so that the accumulated exposure can be determined at a future date. The limit of error must be listed on all instruments used for detection and measurement.

4. An employee who will be exposed to radiation should have an initial physical examination, periodic exams when overexposure has been detected, and an examination when employment terminates. All cases of overexposure and suspected ingestion or inhalation must be immediately referred to a physician. Any loss, theft, damage, or overexposure must be reported to the appropriate authority and the NRC under the requirements of 10 CFR Part 20. Transportation, interstate or intrastate, must comply with the requirements of the Department of Transportation for shipment of explosives and other dangerous articles.

## CONFINED SPACES

According to *The Dictionary of Terms Used in the Safety Profession,* published by the American Society of Safety Engineers,[5] a confined space is a compartment or an area having limited access and (usually) no alternate escape route, having severely limited natural ventilation or an atmosphere containing less than 20 percent oxygen, and having the capacity of accumulating a toxic, flammable, or explosive atmosphere, or of being flooded. Examples include tanks, vaults, ducts, conduits, wells, stacks, tunnels, silos, pits or trenches more than four feet (1.22 meters) deep, ship holds, pipelines, manholes, and sewers. Refer to "Safety Requirements for Working in Tanks and Other Confined Spaces," ANSI Z117.1.

An explosive atmosphere contains a mixture of vapor or gas that is within the explosive or flammable concentration range. The explosive limits are the minimum (lower) and maximum (upper) concentrations of vapor or gas in air or oxygen below or above which explosion or propagation of flame does not occur in the presence of a source of ignition. The explosive or flammable limits usually are expressed in terms of percentage by volume of vapor or gas in air.

A toxic atmosphere is any atmosphere having a toxic or a disease-producing contaminant exceeding the legally established permissible exposure limit or the threshold limit value established by the American Conference of Governmental Industrial Hygienists.

A hazardous atmosphere is any atmosphere having one or more of the above characteristics.

On construction projects, explosive atmospheres can arise in equipment repair shops in winter when doors are closed and a small gasoline spill occurs. In tunnel construction, small seams of coal or oil can leak methane, and only one spark is needed to cause ignition. Many disastrous explosions have resulted when this hazard was not controlled. To avoid hazards associated with confined spaces and explosive and toxic atmospheres, the following measures should be included in all construction safety plans:

1. *Entry procedures:* Whenever it is anticipated that a confined space or an explosive or toxic atmosphere will be encountered, an entry procedure must be written to identify and eliminate or control the hazards. Good references for such procedures are: National Safety Council Data Sheet 1-704-85, "Confined Space Entry Control System for R&D Operations"; the American Petroleum Institute's Recommended Practice, "RP 2015 Cleaning Petroleum Storage Tanks"; NAVFAC Guide Spec 13657; and NIOSH "Criteria for a Recommended Standard—Working in Confined Space." Such procedures should consider such hazards as flammable, toxic, corrosive, or radioactive materials, oxygen deficiency, inadvertent activation of electrical or mechanical equipment, and fire protection systems ($CO_2$ and Halon).

2. *Permit system:* Prior to entry of any personnel into a confined space or into an explosive or a toxic atmosphere, a permit should be issued for each person and location, which should be renewed at the beginning of each shift. Permits (initial and renewed) must be posted at all access locations. Such permits should include, but not be limited to, location of the work, description of the work, employees assigned, entry date and time, isolation checklists, hazards expected, fire safety precautions, personnel safety, results of atmospheric tests performed and who performed them, name of party issuing the permit, and the expiration time.

3. *Testing of atmosphere:* No one is allowed to enter a confined space where a known explosive or oxygen-enriched atmosphere exists. Before entry into a confined space, the work environment must be tested by a competent person using properly calibrated, approved equipment to determine the extent of potential hazards. If the atmosphere cannot be determined by testing and you do not know what is present, the location should be identified as "Immediately Dangerous to Life and Health." (IDLH). The test should consider not only the oxygen content but whether the substance can undergo a change of content (burning styrofoam can create sufficient heat to cause the formation of phosgene gas). While people are working in confined or contaminated areas, testing should be performed on a

continuous and frequent basis. The entry procedure should stipulate the frequency.

4. *Oxygen-deficient atmospheres:* No person should be permitted in atmospheres containing less than 20 percent oxygen without wearing a self-contained breathing apparatus (SCBA) of an approved type or a combination pressure-demand air-line respirator with an integral auxiliary self-contained air supply of at least fifteen minutes' capacity. When ventilation of confined spaces is insufficient, persons must be protected by air-line respirators or SCBAs.

5. *Ventilation systems and respiratory and other equipment:* Mechanical ventilation sufficient to maintain a nonhazardous atmosphere should be provided. If this is not possible, proper respiratory equipment, protective clothing, and a safety line or a body harness should be supplied. Entry into an oxygen-deficient space should not be made without the use of a self-contained breathing apparatus (SCBA), positive-pressure system to prevent entry of outside air into the respirator, or air-line respirator equipped with emergency escape air supply (see respirator protection discussion below). Use of protective clothing and respiratory protection should not be substituted for cleaning and ventilating of spaces prior to entry.

6. *Work in teams/standby person:* Qualified and trained personnel working in confined or contaminated areas should work in teams. One team member should be designated as the standby person, who must remain outside the unsafe area in a safe place and have no other conflicting duties. The standby person must be first-aid and CPR–qualified and well versed in the potential hazards; must be able to communicate (voice, radio, telephone, or signal) with those working in the confined space or contaminated area and with others to obtain assistance; and, in the event of an emergency, should not enter the unsafe area unless equipped with proper rescue equipment and until a new standby person arrives. Appropriate rescue equipment should be located so that it is immediately available to the standby person. It should be tested to ensure that it will fit through the entryway. This test also increases workers' awareness of the difficulty of rescue operations. Communications systems should be tested. The backup standby person should be predesignated and available in the general area. Consideration must be given to the standby person's ability to physically remove incapacitated personnel. A hoist or some other mechanical device might be needed for personnel removal. Supervisors also should allow any personnel with cardiac or respiratory problems to be designated as standby persons.

7. *Blank off or disconnect pipelines:* Precautions should be taken so that no

other liquids, gases, or solids can inadvertently enter the confined space during occupancy. Closing of valves on connecting pipelines is not enough. All pipes should be blanked off or disconnected prior to entry.

8. *Clearance procedures and other controls:* Safe clearance procedures should be developed for deenergizing electrical systems and deactivating machinery, pressure systems, and rotating equipment, as they can be a source of ignition. If welding or cutting is to be performed in a confined space, local exhaust ventilation, other safety equipment, and additional fire protection may be necessary. If paints, thinners, or other hazardous materials are to be used in a confined space, the safety and health information from the Material Safety Data Sheet should be incorporated into the entry procedures and permits. Only explosion-proof lighting/equipment should be used unless atmospheric tests have proved the space to be nonexplosive. The nozzles of hoses and lines for air, inert gas, and steam, when used in the cleaning or the ventilation of tanks and vessels that contain explosive concentrations of flammable gases or vapors, should be bonded to the tank or vessel shell. Bonding devices should not be attached or detached in the presence of hazardous concentrations of flammable gases or vapors.

## RESPIRATORY PROTECTION

The following basic guidelines for respiratory protection should be followed:

1. *Standard operating procedures:* When respirators are to be used, written standard operating procedures (SOP) should be developed that include information on proper use, training of respirator wearers, respirator sealing tests, issuance of respirators, inspection of respirators prior to use, monitoring respirator use, monitoring respiratory hazards, and planning for routine, nonroutine, emergency, and rescue uses of respirators. The SOP should include specific requirements of ANSI Z88.2.

2. *Respiratory protection program:* The construction manager should designate a competent person to develop, implement, and manage the respiratory protection program, which must include the SOP mentioned above and the following requirements:

   a. Respiratory protective equipment must be provided. It should be selected, fit-tested, used, maintained, and stored in accordance with the requirements contained in this chapter, the latest edition of ANSI Z88.2, and the joint NIOSH/OSHA "Standard Completion Program Respirator Decision Logic." In the event of conflicts between these and other referenced standards or regulations, the more stringent should prevail.

b. Personnel must be trained in accordance with the requirements listed above.

c. The medical status of individuals who are to wear respirators should be evaluated, and a statement from a qualified physician should be provided indicating that the individual is medically and physically qualified to wear a specified type of respirator. A respirator equipped with a facepiece as defined in ANSI Z88.2 should not be worn if facial hair comes between the sealing surface of the facepiece and the face, or if facial hair interferes with the valve function. The wearer of a respirator equipped with a full facepiece, helmet, hood, or suit must not be allowed to wear contact lenses. If a spectacle, goggle, face-shield, or welding helmet must be worn with a facepiece, it must be worn so as not to adversely affect the seal of the facepiece to the face.

d. Respirators should not be substituted for engineering or environmental control methods without approval of the construction manager. Only "approved" respiratory protective devices should be provided and used to protect against exposure to hazardous atmospheres containing dusts, fibers, fumes, mists, gases, vapors, or other contaminants, or atmospheres deficient in oxygen or otherwise considered IDLH. "Approved" means that the respirator assembly (the respirator and all of its component parts) has been tested and listed as satisfactory by the Mine Safety and Health Administration (MSHA) and NIOSH. The respirator assembly approval number, criteria for use, and limitations will appear on the respirator assembly and/or its container.

3. *Evaluation of respiratory equipment:* A competent person, knowledgeable about inhalation hazards and respiratory protective equipment, should conduct a step-by-step evaluation to ensure that only respiratory protection appropriate for the conditions of exposure is utilized.

a. Respiratory devices using compressed oxygen, liquid air, liquid oxygen, or mixed gases are prohibited unless approved prior to use by the construction manager. Compressed gaseous or liquid oxygen should meet the requirement of the "U.S. Pharamacopoeia for Medical or Breathing Oxygen." Compressed air should meet at least the requirements of the specification for Type I, Grade D breathing air, and liquid air should meet at least the requirements of Type II, Grade B breathing air as described in the Compressed Gas Association's

(CGA), "Commodity Specification for Air," G-7.1 or the latest edition. Further details on sources of compressed air and its safe use can be found in CGA's Pamphlet G-7, "Compressed Air for Human Respiration," and ANSI Z88.2. Compressed gaseous oxygen should not be used in supplied air respirators or in open-circuit-type, self-contained breathing apparatus that previously used compressed air.

b. Breathing air may be supplied to a respirator from cylinders or air compressors. Cylinders should be tested and maintained in accordance with the applicable Department of Transportation specification for shipping containers, 49 CFR Parts 173 and 178. Cylinders should be marked in accordance with ANSI Z48.1, Federal Specification BB-A-1034a, or interim Federal Specification GG-B-00675A. A compressor used to supply breathing air should be constructed and situated to avoid entry of contaminated air into the air supply system. Breathing-air-type compressors should be designed specifically for intended use and equipped with an approved regulator, a suitable in-line air-purifying sorbent bed, and a filter inserted into the supply line to assure breathing air quality.

c. Oil-lubricated compressors should be equipped with a continuous-monitoring alarm system for high temperature, equipment failure, and carbon monoxide. All alarm systems should be so designed that persons using the respirators are made aware of any change in circumstances and thus can get out immediately in the event of equipment failure. All systems should be calibrated on a daily basis and a record maintained of the results. All oil-lubricated compressors should be equipped with required operational safety devices and reservoirs or equivalent devices, with sufficient capacity to enable the respirator wearer to escape from a contaminated atmosphere in case of equipment failure. When used with a continuous-flow-type air-line respirator, the compressor should provide at least four cubic feet of air per minute (115 liters) for tight-fitting facepieces and at least six cubic feet of air per minute (170 liters) for loose-fitting types such as helmets, hoods, and suits. This requirement is for each user supplied from the compressor.

d. When air-line respirators are used, the supplied air source must not be able to be extended, and the hose length cannot exceed 300 feet from source to user.

4. *IDLH atmospheres:* Employees entering or working in IDLH atmospheres should wear a positive-pressure pressure-demand self-contained compressed air breathing apparatus of an approved type or a combination

pressure-demand air-line respirator with an integral auxiliary self-contained air supply of at least fifteen minutes' capacity. No employee should enter such an IDLH atmosphere unless accompanied by another adequately protected employee or unless wearing a safety line and safety harness tended by a person in a safe area who has no other duties and has the proper equipment available to assist the respirator wearer(s) in case of emergency. An effective signal system that will provide for the quick removal of an incapacitated person should be established.

5. *Breathing-air couplings:* Breathing-air couplings should be incompatible with outlets for nonrespirable air or other gas systems to prevent inadvertent servicing of air-line respirators with nonrespirable gases. Prior to their initial use, the quality of the breathing air supplied from cylinders or compressors should be tested for harmful contaminants in accordance with CGA's "Commodity Specification for Air," G-7.1. A copy of the certificate of analysis from a qualified source showing that the breathing air meets the minimum acceptable criteria should be provided to the construction manager. The frequency of such testing should be, as a minimum, in accordance with ANSI Z88.2 and CGA's G-7.1 or as deemed necessary by the user and/or the construction manager.

6. *Canister and cartridge-type air-purifying respirators:* These types of respirators may be used only under the following circumstances:
   a. The identity and concentration of the contaminant are known and have adequate warning instructions.
   b. The oxygen content of air is at least 20 percent.
   c. There is periodic monitoring of the work area.
   d. The respirator assembly is approved for protection against the specific contaminants and concentration level to be encountered.
   e. The type and the model of the respirator that will be used have been successfully fit-tested on the wearer.

7. *Self-contained breathing apparatus (SCBA):* Supplied air or SCBA should be worn when concentrations of contaminants in the atmosphere are unknown or exceed limitations specified for particular canister/cartridge respirators, or when an abrasive blasting or spraying is being done and high concentrations of harmful substances are present. Only an approved SCBA should be used.

8. *Self-rescuer device:* Employees and others exposed to underground environments such as tunnels or shafts should be provided with an approved self-rescuer device. Employees should have satisfactorily completed an MSHA-approved self-rescuer course or equivalent training prior to going underground.

9. *Welding, cutting, or heating of toxic metals:* These activities in any enclosed space should be performed with general mechanical ventilation, air-line respirators, or local exhaust ventilation sufficient to reduce welding fumes and smoke to safe limits. Local exhaust ventilation will consist of movable hoods placed close to the work to remove smoke and fumes at the source. The safe limits are specified in the latest edition of *Threshold Limit Values and Biological Exposure Indices* of the American Conference of Governmental Industrial Hygienists (ACGIH). Toxic metals most commonly encountered are:
   a. Zinc-bearing or filler metals or metals coated with zinc-bearing materials (galvanized).
   b. Cadmium-bearing metals or metals coated with cadmium-bearing materials.
   c. Chromium-bearing metals or metals coated with chromium-bearing materials.
   d. Metals coated with mercury-bearing metals.
   When welding, cutting, or heating is done that involves beryllium-containing base or filler metals or metals containing lead, other than as an impurity, or metals coated with lead-bearing materials, air-line respirators should be used, except when such work is done in an enclosed space. In this case both local exhaust ventilation and air-line respirators are required. Other persons exposed to the same atmosphere as the welders or the burners should be protected in the same manner as the welders or the burners. Welding, cutting, and heating not involving conditions or materials described in this section normally may be done without mechanical ventilation or respiratory protective equipment; however, where, because of unusual physical or atmospheric conditions, an unsafe accumulation of contaminants exists, suitable mechanical ventilation or respiratory protective equipment should be provided.

## PHYSICAL CONTACT

Because spills of toxic substances can not only contaminate the air but result in skin contact through saturated clothing, protective clothing should be worn to avoid this hazard. Exposure of persons to skin absorption or ingestion of any chemical or physical agent above those specified concentrations in ACGIH's *TLV Handbook* should be avoided. In the event of conflicts between these and other cited standards or regulations, the more stringent should prevail. The construction manager should ensure compliance with all applicable standards and regulations to reduce contaminant concentration levels to as low as reasonably achievable (ALARA). Ingestion of toxic materials most often is traced to contamination of snacks or lunches by either dust or vapors. Construction planning should provide designated lunchrooms far removed from work areas where contamination might be a problem. Liquid caustics should be stored in caustic-resistant drums (plastic) where they cannot be damaged, as a leaking caustic substance from a damaged drum can cause severe burns to the feet of an unsuspecting worker, who steps into it only to find that the best leather work shoes are no match for the caustic liquid.

## REFERENCES

1. *The Toxic Substance List,* U.S. Department of Health and Human Services, Public Health Service, Centers for Disease Control, National Institute for Occupational Safety and Health.
2. Sax, N. Irving, *Dangerous Properties of Industrial Materials,* Van Nostrand Reinhold, New York, 1994.
3. *Code of Federal Regulations,* CFR29, Subpart Z, 1910.1000 through 1910.1500, U.S. Department of Labor, OSHA.
4. *Safety and Health Requirements Manual,* EM 385-1-1, U.S. Army Corps of Engineers.
5. Abercrombie, Stanley A., Editor, *The Dictionary of Terms Used in the Safety Profession,* 3rd Edition, American Society of Safety Engineers, Des Plaines, IL, 1988.

# Chapter *17*

## *Planning Safety for People*

This world is given as the prize for men in earnest.
(Fredric Robertson, *Forbes* magazine)

In a practical sense planning is a social contract that brings about a partnering of the various contractors and suppliers of construction services and develops an informal type of relationship among all parties, management and workers alike, in which they all understand that their ideas and thoughts are of value and that a safe and productive workplace will be their reward. In a very short period of time, since the building of the Panama Canal, we have switched from the manual movement of dirt by shovel to bigger and more efficient machines that can excavate, transport, and deposit dirt by the cubic yard, steadily decreasing the hours of human labor required and the cost of moving the earth. As each of the new mechanical wonders has been introduced, workers have become more and more skilled in using them; and their daily experience with these mechanical monsters has made operators extremely perceptive about the type of training, design improvements, and methods of use needed to make the equipment safer and more productive. This evolution from coolie-type labor to a highly sophisticated workforce has been accomplished by an informal partnering and sharing of ideas, not by voluminous paperwork or an oppressive management hierarchy.

Two factors are significant in maximizing workers' performance:

1. They must be competent and fit to do what is required of them.
2. They must be provided a safe workplace to do their jobs.

An unsafe workplace compromises workers' performance when lives are put in jeopardy, as people who are aware of a hazard and must constantly avoid it are put in a stressful situation that reduces their productivity. Likewise, if they are not aware of a hazard and are injured by it, or someone else is, they feel abused and cheated, and the morale of the entire work crew is lowered. Even in war, firepower is known to be more effective than heroics, as the needless loss of life usually does not lead to victory. When management recognizes that hazards produce stress and compromise workers' performance, it will see hazard prevention as a money-maker. Many craftspeople, foremen, supervisors, and managers are not able to recognize hazardous circumstances because they have not had special training in hazard recognition or do not hold appropriate licenses, certificates, or other endorsements to give them the needed authority to assure the safe conduct of the work as defined in the safety plan.

## FITNESS

Physical and mental fitness enables people to perform their work safely, even when reasonable safeguards have been provided, and plays a vital role in the success of any project. All prospective workers should be screened before starting work. Just as it would be unreasonable to allow the blind to drive automobiles by assuming that they can be responsive to the commands of the seeing, so must we guard against bringing the unfit onto the job site. Odd as it may seem, a recorded instance exists where a legally blind man was found to be operating a crane.

## APPROPRIATE PROFESSIONAL REGISTRATION OR CERTIFICATION

One of the objectives of technical registration by the various states is to ensure that engineers are qualified and consider safety as a top priority. Construction safety plans also should include a provision requiring everyone to submit proof of appropriate professional and craft registration or certification. Construction managers should be licensed professional engineers. Likewise, the safety engineer should be a Certified Safety Professional (C.S.P.) as well as a professional engineer. The Board of Certified Safety Professionals (BCSP), which has been established for over twenty-five years and has over 8,000 Certified Safety Professionals in practice, now will quantify the performance of construction safety professionals, site safety specialists, and first-line supervisors and provide certification for these three groups. For additional information, contact the BCSP at 208 Burwash Avenue, Savoy, IL 61874, telephone (217) 359-9263. All engineers responsible for preparing shop drawings, foundation specifications, and structural analysis should be required to stamp/seal these documents with their professional seals. If a registered engineer does not feel comfortable with any of the plans, registration gives the engineer the authority *not* to stamp and sign them, making such plans worthless until changed.

## LICENSING OF EQUIPMENT OPERATORS

In-house licensing of equipment operators is very important to prevent inexperienced and unqualified operators from working on the job site. ANSI and other standards contain guidelines for hiring crane operators, and these guidelines can be applied to operators of other types of equipment. Currently Specialized Carriers & Rigging Association, 2750 Prosperity Avenue, Suite 620, Fairfax, VA 22031-4312, telephone (703) 698-0291, is developing a nationwide program for crane operator certification. Before giving anyone authorization to operate equipment or machinery on a project, examine and document each operator's qualifications to include, but not be limited to, the following:

1. Summary of operating experience.
2. List of previous injuries or property damage resulting from past personal operation of equipment or machinery.
3. Education.
4. Doctor's current physical examination for:
   a. Vision with a minimum of 20/30 Snellen in one eye and 20/50 in the other, with or without corrective lenses; and the ability to distinguish colors.
   b. Adequate hearing, with or without a hearing aid.

    c. Sufficient strength, endurance, agility, coordination, and speed of reaction to meet equipment needs.

    d. Absence of physical defects or emotional instability that could interfere with performance.

    e. Any proneness to seizures or loss of physical control.

    f. Any evidence of drug usage.

5. Passing performance and oral tests on equipment or machinery to which assigned, including:

    a. Demonstration of various modes of operation.

    b. Demonstration of ability to attach and detach accessories.

    c. Knowledge of function of each control and accessory.

    d. Demonstration of ability to handle equipment or machinery safely in all work circumstances.

    e. Knowledge of all safety requirements applicable to equipment or machinery.

    f. Knowledge of hazards inherent in the operation of the equipment.

    g. Knowledge of procedures to follow when unsafe circumstances have been discovered.

6. Passing an appropriate written test on the safe use and operation of the equipment.

Equipment operators depend upon signalers or flaggers to direct them. One of the most flagrant abuses of job-site safety occurs when a crane operator has to work without someone guiding the load and giving proper signals. Signalers and flaggers should not be given other distracting assignments. As the sole communicator to an operator, they should also have adequate physical faculties to do the job, knowledge of appropriate signals, and an awareness of inherent hazards.

## OTHER EQUIPMENT: LICENSING AND CERTIFICATION OF AIRCRAFT, BOATS, AND OTHER MARINE EQUIPMENT

Aircraft, boats, and other marine equipment require Federal Aviation Administration and U.S. Coast Guard certification, and appropriate licenses and certifications need to be a condition of employment when such equipment is used.

Do not rely totally upon a current state driver's license for the operation of company vehicles and other equipment. Check with the Motor Vehicle Department to determine if the driver has a record of collisions, traffic citations, and DWI arrests. Unsuspected alcohol and drug abuse often have led to catastrophe.

## QUALIFICATIONS OF EXAMINERS

OSHA requirements often refer to the need for a competent or qualified person to examine and determine the safety of equipment, machinery, structures, or facilities. To meet this requirement, the following should be considered:

1. Work experience.
2. Appropriate professional licenses, certifications, or registrations.
3. Special OSHA or other academic training.

## TASK EXAMINATION

The actual tasks people will perform must be examined in the planning process, as far too often tasks exceed a worker's ability. For example, back injury occurs very frequently on construction projects, so any item that must be lifted or carried by a worker should not exceed 45 pounds. The old 200-pound railroad nail keg is no longer considered reasonable, and most nails now come in approximately 50-pound cardboard boxes. Slips and falls on unsafe accessways cause many broken bones and sprains, and tasks that cause a worker to overreach should be avoided.

An overlooked condition affecting task performance is the mental stress arising when workers are assigned responsibilities that cannot be met and when they are held accountable for reaching unattainable goals. Safety planning always must consider the limitations of the equipment or the machinery that the worker must deal with, and the inability of people to recognize hazardous conditions. On the other hand, a worker's input into safety planning can be invaluable, as the worker may be able to suggest a safer and more productive solution. Giving workers the opportunity to be partners with management develops a stress-free team approach. Do not put the worker who asks "why" on the spot, as there may be a very good reason for such a question, either a lack of understanding about the task or apprehension because a hazard the worker perceives is not included in the planning. When there are no questions, be wary; for psychologists say that both children and adults are inclined to keep quiet because they have learned that when they ask questions, they often are ridiculed by their peers. When reviewing safety plans with work crews, be sure to give them a chance to ask questions; and if the response is still dead silence, try, in a nonintimidating manner, to ask them how they think the task should be accomplished. The few minutes spent in providing an opportunity for worker participation, along with being flexible enough to improve the plan by incorporating workers' suggestions, can pay big dividends, not only in providing a safer workplace but in developing workers' confidence and pride in their work.

## PERMITS

Specific high-hazard operations need to be controlled by the issuance of permits restricting who will do the work, when and how it will be done, and under whose supervision. Additional information on permits may be found in Chapter 21.

## OTHER CONSIDERATIONS

What people do or might do should be taken into consideration in planning a project. People overqualified for a position create one set of risks, those underqualified create different risks, and a common thread uniting both of these types and creating another set of risks[1] is boredom from performing repetitious, automated tasks. Planning should create a challenging work environment, offering both personal achievement and reward.

## REFERENCE

1. Gross, Vernon L., *Managing Risk,* Prentice-Hall, Englewood Cliffs, NJ, 1987.

# Chapter 18

*Training*

A driving ambition is of little use if you are on the wrong road.

(Frank Tyger)

If excellence in safety performance is to be achieved, an ongoing in-house safety education program must be part of the construction safety plan, as its inclusion will pay big dividends in the long run.[1] In construction injury litigation, liability often is hard to avoid if it can be shown that the construction manager did not have adequate safety training. It is a recipe for disaster for management to fail to recognize its own need for training and its responsibility to oversee the training of its subordinates.

There is no shortage of packaged safety training programs and films, but the vast majority focus on training the worker to live with and accommodate hazards, and do not address what can be done to eliminate or control them. Such an approach does little to encourage worker participation in safety or to inspire faith in the safety plan, as most workers already know that many of the hazardous circumstances they must deal with can be resolved by using safer construction methods, safer equipment, and available safeguards. The construction manager should make sure that the safety engineer screens all safety training programs to see that they address hazards and their prevention and that they do not characterize workers as stupid or ridicule them for alleged unsafe behavior.

Training should emphasize the hazards workers may encounter on the job so that task planning can include necessary changes to eliminate dangerous circumstances. Management, by its actions, must convey its support to workers; this is done by introducing safer methods, safer equipment, and the use of safeguards. The bottom line is that you lose your investment in training dollars when hazard prevention is ignored and workers are required to work in an unsafe workplace, which causes conflict and low morale that severely limit productivity.

Some progressive safety engineers have made three- to five-minute on-the-job videos that show hazardous circumstances and how they can be prevented, and they use them at toolbox safety meetings. Some have even gone so far as to provide a portable videocassette player that can be operated from the seat of a pickup truck with a twelve-volt DC adapter so that foremen can present explicit and memory-retainable information to work crews. When the safety suggestions of foremen and their crews are recorded on videotape for training purposes, their involvement in safety planning becomes meaningful and leads to future improvements.

The construction management team and those designers who will visit the project should be trained to recognize the inherent hazards and pertinent provisions of OSHA, ANSI, and other standards to be able to identify and report any dangerous circumstances. (See Chapters 20 and 33.)

This book does not address development and presentation of training programs, as many textbooks have been published by competent authors on effective training methods. Their objective is to be sure that safety plans include necessary safety training for all personnel, including the project manager.

## REFERENCE

1. Brauer, Roger L., *Safety and Health for Engineers,* Van Nostrand Reinhold, New York, 1990.

# Chapter 19

## *Monthly Management Safety Oversight Meetings*

Who works, achieves and who sows, reaps.

(Arab proverb)

Once a contract has been awarded, it is exceedingly important that the safety program as outlined and agreed upon at the pre-notice-to-proceed safety conference is kept on target, and that all mileposts on the Critical Path Method charts are reached. This is accomplished by holding a monthly safety oversight meeting chaired by the construction manager with the assistance of the safety engineer and attended by superintendents of all major contractors and subcontractors—and, when high-risk hazards are anticipated, any of their foremen who have direct contact with current work. If high-risk hazards involving utilities, government agencies, or suppliers of goods and services are foreseen, the construction manager should request the attendance of their representatives by letter so that their expertise and support are available. Attendance at these monthly meetings should be mandatory.

At this meeting the construction manager should call for a review of progress made during the previous month and a discussion of any deficiencies or hazardous circumstances that have become apparent since the previous meeting, as well as any action taken. The next month's safety plans and anticipated high-hazard activities then should be reviewed and appropriate preventive measures outlined.

The safety engineer should prepare a detailed outline of topics that will be discussed so that all safety problem areas can surface for discussion. Prior to the meeting, the safety engineer should poll all superintendents of the major contractors and subcontractors for hazard topics pertinent to their activities, should be familiar with current workplace conditions, and should gather supporting information to include, but not be limited to, injury data, reference materials, pertinent standards, and the impact of the hazard in terms of potential injury, property damage, or regulatory government citation.

Minutes of the meeting should be taken. A written record should be kept of the resolution of all hazardous circumstances and the agreed-upon hazard prevention measures, recording who will do what and when. Each attendee should be given a copy, and a copy should be retained in the construction manager's files. This type of recordkeeping is invaluable in the event that claims involving costs or injuries are made at a later date.

The real worth of monthly management safety oversight meetings is in the lives they save. Be sure these meetings are included in planning, as they are critical in providing a safe operation. Lessons learned in these meetings should be recorded for inclusion in phase planning or for future projects.

# Chapter *20*

## *Project Monitoring*

Ignorance of the law is not an excuse.

(Common law)

The biggest deterrent to a safe construction workplace is the negative mind-set that hazard prevention is someone else's business. To have a thousand eyes looking for hazardous circumstances on a construction project is a construction manager greatest asset, and it is very achievable if everyone connected with the project is collaterally responsible for safety.

The construction inspector who is primarily concerned that pipe joints in a trench are properly connected and rest on proper bedding materials on a proper grade should also be seeing whether the shoring of the trench is adequate, whether there are powerlines above the trench that must be avoided by the excavator, and whether, instead of there being an open hook that can easily dump straps, a ring is welded to the excavator's bucket to shackle the straps.

The designer of mechanical equipment who is visiting the site during installation should be sure that floor openings where the equipment will be installed remain covered until just before installation to avoid the serious hazard of falls through uncovered floor or roof openings.

Construction managers should ask the landowner and all suppliers of goods and services who visit the job site to look for hazards and report them immediately. In construction personal injury litigation, failure to initiate reasonable care to overcome hazardous circumstances is indefensible. It takes only minimal orientation to develop sensitivity to what constitutes a life-threatening circumstance.

All construction engineers and supervisors should be required to have a basic knowledge of inherent hazards and of pertinent safety requirements of OSHA, ANSI, and other standards. (See Chapter 18.)

The construction manager should see that the safety engineer maintains a library of current and pertinent requirements and standards that is accessible to the entire staff.

Probably one of the most frequent sources of trouble is the machinery or equipment brought onto a project. Often it has not been repaired as stated, or it has been repaired, serviced, or modified in a haphazard, improper, and incomplete manner. It can pay big dividends for the project safety engineer to check the validity of repair records.

Because construction managers and project safety engineers sometimes must fight one "brush fire" after another, it is easy to overlook what might seem to be a small irregularity. Sometimes small irregularities are like the tip of an iceberg, indicating a much larger problem. The thing to do is to make a note of any irregularity, bring it up at the monthly management safety oversight meetings (see Chapter 19), and assign someone to investigate it. Usually the investigation will show that the operating-level workers knew about the irregularity and had a very good explanation for it. Laborers and first-line supervisors, when confronted with a problem, usually have some very practical solutions for it that should be considered. Members of management do not pay enough attention to workers' input or, more important, do not encourage their participation in making the workplace safer. The solution is to make the safety office accessible to all the workers and to encourage them to call in, leave a note, or just ask for help in solving a safety problem. It is not always what you see in a "walk-around" but what you hear. Build the confidence of your crews by being a good listener. This will open the door to much good information. If changes are necessary, make them a team effort, giving recognition to those workers who brought the problem to light. When people are made to feel they are part of the monitoring program for safety, they will be more open to alerting the safety office when a hazardous condition is suspected. This is the optimum in project safety monitoring because everyone is doing it. Do not shoot from the hip with immediate answers, but take the time to evaluate all options and their impact, and, most important, be sure everyone involved is informed of the changes.

As discussed in Chapter 11, the first item in the construction safety plan is to establish a fast-track chain-of-command for the immediate resolution of all hazardous circumstances as they are identified. There is nothing worse than someone's being injured or killed because an identified hazardous circumstance was not controlled promptly.

# Chapter *21*

## *Phase Planning*

To business that we love we rise betime,
And go to 't with delight.

(Shakespeare, *Antony and Cleopatra,* IV, 4)

After a project's master construction safety plan has been prepared, each major phase of construction must be scrutinized for safety. Each task within each phase should reviewed for any hazards that might arise and to see what can be done to prevent them from causing injury and property damage. The next ten chapters (Chapters 22–31) cover safety planning for each major phase of construction, from site preparation, excavating, concrete formwork and placement, steel erection, mechanical installation, electrical systems, interiors, exteriors, and roofing to maintenance. Where possible, this phase planning should be integrated with CPM charts. Superintendents or foremen who oversee a particular phase of construction should prepare a written phase safety plan with hands-on assistance from the project safety engineer, to be reviewed and approved by the construction manager. (See Chapter 8, dealing with duties of the construction manager and the project safety engineer.) To make a thorough phase safety plan, each work task should be examined individually and a job safety analysis worksheet prepared as described in the National Safety Council's *Accident Prevention Manual for Industrial Operations, Administration and Programs.* [1]

The phase safety plan must address all hazards that might arise during a particular phase of work and must outline specific safeguards, procedures, and permits that will be required to minimize the risk of injury and property damage. When high-risk hazards are encountered, sufficient redundancy must be incorporated so that several safeguards and procedures must be simultaneously violated before a hazard can cause harm.

The phase safety plan should include the following:

1. A list of activities—excavation, tunneling, highway paving, concrete work, wood framing, steel erection, bridge building, and so on.
2. Permits needed. Where high risk is involved in a task, a permit system must be set up so that work can proceed in a safe manner. Permits should state who is to be involved (only those qualified should be assigned to do the task), the time frame in which the work is to be done, where it should be done so that it does not conflict with other activities, work methods and procedures to be used, and so on. Permits also must outline specific hazard prevention measures. For example, the moving of a crane from one location to another should be done by permit to avoid injury or property damage from contact with any powerlines that might be strung in the path of travel and within reach of the boom; and disassembling a latticework boom should be done by permit to avoid unintended collapse when insufficient cribbing is provided for support. Blasting is another type of work that should be contolled by permit, not only to protect workers but to avoid damage to surrounding facilities. The project safety engineer needs to examine high-risk activities in each phase and identify which tasks require permits, which will be issued and authorized by the construction manager.
3. All job safety analyses made for individual work tasks.
4. A list of equipment, machines, and tools to be used. When possible, identify by make, model, date of manufacture, inherent hazards, and needed safety accessories.
5. A list of physical and health criteria, experience, and training required for personnel.
6. A list of special operating procedures or methods.
7. Designation of the chain of command for authority.
8. Designation of a phase safety officer.
9. A list of all contractors, subcontractors, suppliers, and others who will be involved in this phase.
10. Schedule of safety meetings and toolbox topics.
11. Schedule of phase safety inspections and performance audits.
12. Instructions for reporting hazards and injuries.

Each of the following ten chapters discusses in detail some of the hazards to look for in each phase, the prevention measures that will need to be considered, and the applicable requirements that must be incorporated into the phase safety plan.

Many of the guidelines in some of the previous chapters and in the next ten chapters have been summarized and condensed from the U.S. Army Corps of Engineers' *Safety and Health Requirements Manual,* EM 385-1-1.[2] The requirements in this manual always meet OSHA standards and often exceed them, and, when a hazard is exceedingly dangerous, they suggest even more stringent standards of care. It should be noted that specific requirements are listed as measures that "should" rather than "shall" be initiated or taken as specified in OSHA or other references because this book provides guidance only on the avoidance of hazardous circumstances through planning and is not a requirements manual.

Nearly every type of phase planning includes the use of cranes. Appendix 2 is a generalized critical lift plan for the safe use of cranes.

The next ten chapters should be considered as checklists of pertinent do-and-don't requirements to avoid specific hazards encountered in the particular phase being examined. As management consultant Tom Peters has remarked, "You don't know where you're headed until you begin to march."[3] The sequence he respects the most is: "Ready. Fire. Aim." Do not let planning become a ritual rain dance that never brings any rain (or makes the job any safer). Always be aware of what has happened on other projects, and in your phase planning, in addition to these checklists, target hazards that have caused other similar projects unnecessary injury, death, and property damage.

## REFERENCES

1. National Safety Council, *Accident Prevention Manual for Industrial Operations,* Vol. 1, *Administration and Programs.*
2. U.S. Army Corps of Engineers, *Safety and Health Requirements Manual,* EM 385-1-1.
3. Peters, Tom, "Management," syndicated newspaper column, Tribune Media Services.

# Chapter *22*

## *Phase Planning for Site Preparation*

At King Solomon's command, they cut fine large stones for the foundation of the temple.*

(1 Kings 5:17)

The phase safety plan for site preparation must cover all activities that need to be done within that phase of work and must list basic precautions to be taken, covering such topics as removal of powerlines where high-clearance equipment will be used, land clearing, demolition, drilling, blasting, building access, roadways, and so on.

## SAFETY PLANNING FOR POWERLINES

The first thing that should be done at a project site is the arrange for the removal of powerlines from the work areas where cranes and other high-clearance equipment will be used. Contractors are required to inform the

---

*Biblical quotations in this book are taken from the *Good News Bible*, Today's English Version, published by the American Bible Society, New York.

local electric utility that cranes will be used in specific areas of the work site so that the utility can either bury temporary or permanent powerlines or move them to other locations where such equipment will not be used. The electric utility is the ultimate authority for the placement of powerlines outside of harm's way. This is a very important part of safety planning.

Cranes cannot be safely operated near powerlines. The crane operator, persons guiding the load, and those closely involved in the operation need some visual guidance on the ground to make them aware of the danger zone so that they can conduct their work outside of it. The best approach is to institute a mapping system.[1] The area within a radius of ten feet in any direction from powerlines is an unsafe area and must be clearly marked off on the ground by tape, fences, barriers, or other highly visible markers so that everyone at the worksite can assure that the crane is positioned so the boom or the hoist line cannot intrude into this danger zone during sluing or other operations. Job-site foremen, operating personnel, riggers, crane operators, and signalers must be trained in mapping the danger zone and in placing visible markers on the ground to define this zone. The easiest way to identify the parameters of the danger zone on the ground is to step off fifteen feet from each side of each power pole in the work area and make a continuous line from pole to pole by placing a visual marker on the line so that any encroachment of the hoist line, boom tip, or load into this danger zone will be visible to all personnel involved in the craning operations. (See Figures 14-6 through 14-8.)

If a crane boom, hoist line, or load can penetrate the danger zone and other positioning is impossible, the electric utility must be notified to deenergize, relocate, bury, or insulate the powerlines while the crane is operating in that location. (See OSHA §1910.180.J, "Operating Near Electric Powerlines," and OSHA §1926.550.A.15.) The utility also should be asked to lock out the automatic reclosure system (OSHA §1926.955[e][5]) to avoid reenergizing the powerline in the event of an inadvertent contact.

Various safeguards are employed by some contractors who must do pick-and-carry operations with cranes on highways, roads, and streets where removal or deenergization of the powerlines is not practical have equipped their cranes with proximity alarms as early warning devices, which enable them to alter their work procedures to avoid powerline contact. Another approach is to notify the power company of impending work and request that the lines be covered with line covers. Because powerlines can be camouflaged by many backgrounds, making them almost invisible to the naked eye, and visual judgment of clearances is not always very accurate— which human factors experts have been telling us for years—some sort of early warning system is essential. Another important consideration is that people can focus only on one target at a time; thus crane operators, riggers, and signalers, while focused on guiding a load, may be completely unaware that the boom or hoist line is getting too close to a powerline. Insulated links, which isolate the suspended load from the hoist line and boom, also have proved their worth as a redundant backup safeguard. In Australia the use of nonconductive taglines is required. All these safeguards increase workers' awareness of the inherent dangers of working around powerlines, and there is no record of injury arising from the use of these safeguards when their manufacturers' recommendations are followed. None of these safety devices is considered a substitute for OSHA's required ten-foot clearance.

## SAFETY PLANNING FOR LAND CLEARING

Land clearing and grading can be very hazardous, especially the removal of trees and heavy brush. All tractors, scrapers, graders, and compactors should have rollover protective systems (ROPS) for safe land clearing. Dozers used to clear brush should be provided with heavy-duty brush guards to prevent the intrusion of limbs and branches into the operator's cab. Such brush guards must be certified to meet ROPS requirements.

Cleared, leveled, and compacted ground provides the safest access for all foot traffic.

## SAFETY PLANNING FOR DEMOLITION

The following guidelines should be used for demolition:

1. Before workers are permitted to start demolition, an engineering survey of the structure and the demolition plan should be made by a competent person to determine the condition of the framing, floors, and walls and the possibility of unplanned collapse of any portion of the structure. Any adjacent structure where workers may be exposed should be similarly checked. The construction manager should have evidence in writing that such a survey has been performed.

2. When workers work within a structure to be demolished that has been damaged by fire, flood, explosion, or other cause, the walls or the floor should be shored or braced.

3. All electric, gas, water, steam, sewer, and other service lines should be shut off, capped, or otherwise controlled outside the building line before demolition is started. In each case, any utility company that is involved should be notified in advance.

4. If it is necessary to maintain any power, water, or other utilities during demolition, such lines should be temporarily relocated and protected.

5. It should be determined if any hazardous chemicals, gases, explosives, flammable materials, or dangerous substances have been used in any pipes, tanks, or other equipment on the property. When any such substance is apparent or suspected, testing and purging should eliminate the hazard before demolition is started.

6. All broken glass should be removed.

7. Where a hazard exists to workers of falling through wall openings, the openings should be protected to a height of approximately forty-two inches.

8. All floor openings not used as material drops should be covered over with material substantial enough to support the weight of any load that may be imposed. Such material should be secured to prevent accidental movement.

9. Except for the cutting of holes in floors for chutes or holes through which materials will be dropped, the preparation of storage space, and similar preparatory work, the demolition of exterior walls and floor construction should begin at the top of the structure and proceed downward. Each story of the exterior wall and floor construction should be removed and dropped into the storage space before the removal of exterior walls and floors in the story below is begun.

10. Entrances to multistory structures that are being demolished should be protected by sidewalk sheds or canopies, or both, providing protection from the face of the building for a minimum of eight feet. All such canopies should be at least two feet wider than the building entrances or openings (one foot wider on each side), and should be capable of sustaining a load of 150 pounds per square foot.

11. Mechanical equipment should not be used on floors or working surfaces unless such floors or surfaces are of sufficient strength to support the imposed load.

12. Selective demolition by explosives should be conducted in accordance with the blasting guidelines presented in this chapter.

13. Stairs, passageways, and ladders used in demolition should meet the following guidelines:

    a. Only those stairways, passageways, and ladders designated as means of access to the structure of a building should be used. Other access ways should be closed at all times.

    b. All stairs, passageways, ladders, and incidental equipment covered by this section should be periodically inspected and maintained in a clean and safe condition.

    c. In a multistory building, a stairwell being used should be illuminated by either natural or artificial means and covered over at a point not less than two floors below the floor on which work is being performed. Access to the floor where the work is in progress should be through a lighted, protected, and separate passageway.

14. Debris removal during demolition needs to meet the following requirements:

    a. When debris is dropped through holes in the floors without chutes, the area onto which the material is dropped should be enclosed with barricades not less than forty-two inches high and not less than six feet back from the projected edge of the opening above. Signs warning of the hazard of falling materials should be posted at each level. Removal should not be permitted in this lower area until debris handling ceases above.

    b. All material chutes, or sections thereof, at an angle of more than 45 degrees from the horizontal should be enclosed, except for openings equipped with closures at or about floor level for the insertion of materials. The openings should not exceed forty-eight inches in height measured along the wall of the chute. Such openings, when not in use, should be kept closed at all floors below the top of the floor.

    c. A substantial gate should be installed in each chute at or near the discharge end. A competent worker should be assigned to control the operation of the gate, and the backing and loading of trucks.

    d. When operations are not in progress, the area surrounding the discharge end of a chute should be closed off.

    e. Any chute opening into which personnel dump debris should be protected by a guardrail forty-two inches above the floor or other surface on which personnel stand to dump the material. Any space between the chute and the edge of openings in the floors through which it passes should be covered.

    f. Where the material is dumped from mechanical equipment or wheelbarrows, an attached toeboard or bumper, not less than four inches thick and six inches high, should be provided at each chute opening.

g. Chutes should be designed and constructed of such strength as to eliminate failure due to impact of materials or debris loaded therein.

h. The storage of waste and debris on any floor should not exceed the allowable floor loads.

i. In buildings having wood floor construction, the floor joints may be removed from not more than one floor above grade to provide storage space for debris, provided that falling material is not permitted to endanger the stability of the structure.

j. When wood floor beams serve to brace interior walls or free-standing exterior walls, such beams should be left in place until other support can be installed to replace them.

k. Floor arches, to an elevation of not more than twenty-five feet above grade, may be removed to provide storage area for debris, provided that such removal does not endanger the stability of the structure.

l. Storage space into which material is dumped should be blocked off, except for openings for the removal of materials. Such openings should be kept closed when material is not being removed.

m. Floor openings should have curbs or stop-logs to prevent equipment from running over the edge.

n. Any opening cut in a floor for the disposal of materials should be not longer in size than 25 percent of the aggregate of the total floor area, unless the lateral supports of the removed flooring remain in place. Floors weakened or otherwise made unsafe by demolition should be shored to carry safely the intended imposed load for demolition.

15. Wall removal during demolition should meet the following requirements:

a. Masonry walls, or sections of masonry, should not be permitted to fall upon the floors of the building in such masses as to exceed the safe carrying capacities of the floors.

b. No wall section that is more than one story in height should be permitted to stand without lateral bracing, unless such wall was designed and constructed to stand without such lateral support and is in a condition safe enough to be self-supporting. All walls should be left in a stable condition at the end of each shift.

c. Workers should not be permitted to work on the top of a wall at any time without appropriate fall protection. Even with fall protection, when the weather is bad, no one should work on top of a wall. Structural or load-supporting members on any floor should not be cut or removed until all stories above such a floor have been demolished and removed. This requirement should not prohibit the cutting of floor beams for the disposal of materials or for the installation of equipment, provided that the requirements for debris removal and floor removal are met.

e. Floor openings within ten feet of any wall being demolished should be planked solid except when workers are kept out of the area below.

f. In "skeleton-steel" construction, the steel framing may be left in place during the demolition of masonry. Where this is done, all steel beams, girders, and structural supports should be cleared of all loose material as the masonry demolition progresses downward.

g. Walkways or ladders should be provided to enable workers to reach or leave any scaffold or wall safely.

h. Walls that serve as retaining walls to support earth or adjoining structures should not be demolished until such earth has been braced or adjoining structures have been underpinned.

i. Walls serving to retain debris should not be so used unless capable of safely supporting the imposed load.

16. Floor removal during demolition should meet the following requirements:

a. Openings cut in a floor should extend the full span of the arch between supports.

b. Before any floor arch is demolished, debris and other material should be removed from such arch and other adjacent floor area. Planks not less than two inches by ten inches in cross section, full-sized undressed, should be provided and should be used by workers to stand upon while breaking down floor arches between beams. Such planks should be so located as to provide a safe support for workers should the arch between the beams collapse. Straddle space between planks should not exceed sixteen inches.

c. Safe walkways, not less than eighteen inches wide, formed of wood planks not less than two inches thick or of equivalent strength, should be provided and used by workers when necessary to enable them to reach any point without walking upon exposed beams.

d. Stringers of ample strength should support the flooring planks, and the ends of such stringers should be supported by floor beams or girders and not by floor arches alone.

e. Planks should be laid together over solid bearings with the ends overlapping at least one foot.

f. When floor arches are being removed, employees are not to be allowed in the area directly underneath, and such an area must be barricaded to prevent access to it.

17. Steel removal during demolition needs to meet the following requirements:

    a. When floor arches have been removed, planking will be provided for the workers razing the steel framing.

    b. Steel construction will be dismantled column by column and tier by tier (columns may be in two-story lengths).

    c. Any structural member being dismembered will not be overstressed.

18. Mechanical demolition needs to meet the following requirements:

    a. No person will be permitted in any area that can be affected by demolition when balling or clamming is being performed. Only those persons necessary for the operations shall be permitted in this area at any other time.

    b. The weight of the demolition ball should not exceed 50 percent of the crane's rated load, based on the length of the boom and the maximum angle of operation at which the demolition ball will be used, or it should not exceed 25 percent of the nominal breaking strength of the line by which it is suspended, whichever is less.

    c. The crane boom and the load line should be as short as possible.

    d. The ball will be attached to the load line with a swivel connection to prevent twisting of the load line and will be attached by positive means so that the weight cannot accidentally disconnect.

    e. When walls or portions thereof are pulled over, all steel members affected must have been cut free.

    f. All roof cornices or other ornamental stonework will be removed before walls are pulled over.

    g. During demolitions, continuing inspections by a competent person will detect hazards resulting from weakened or deteriorated floors, walls, or loosened material. No one will be permitted to work where such hazards exist until they are corrected by shoring, bracing, or other means.

    h. Also see Chapter 16 for guidelines to follow when burning/cutting metal to avoid dangerous fumes.

## SAFETY PLANNING FOR DRILLING AND ROTARY-JET PIERCING

Drilling and rotary-jet piercing during site preparation should conform to the following guidelines:

1. All drill holes should be of greater diameter than the cartridges of explosive materials used.
2. Drilling should not be done in an area already blasted until remaining "bootlegs" (holes that do not detonate full depth) are examined for unexploded charges and the total area has been examined to make sure that there are no unexploded charges remaining. Never insert a drill, pick, or bar into bootlegs, even if examination fails to disclose explosives. When misfires have occurred, and drilling must be done in an area where undetonated holes may exist, holes should not be drilled where there is danger of intersecting a misfired hole. All drilling necessary to neutralize misfires must be done under the supervision of a competent person who has a working knowledge of the explosive materials involved and is familiar with the conditions under which the misfired holes were drilled, loaded, primed, and initiated.
3. Drilling and loading operations should not be carried on in the same area. Drilling should be separated from loaded holes by at least the depth of the loaded hole but in no case by less than fifty feet.
4. Dust from any drilling should be controlled by wet drilling or other means that provide similar control.
5. Workers should not be on a mast while the drill-bit is in operation unless they are provided with a safe platform from which to work, and they are required to use safety belts to avoid falling.
6. Drilling crews and others should stay clear of augers or drill stems that are in motion. Persons should not pass under or step over a moving stem or auger.
7. When a drill is being moved from one drilling area to another, drill steel, tools, and other equipment should be secured and the mast placed in a safe position. Particular care should be taken when a drill is being moved under or around overhead powerlines.
8. In the event of power failure, drill controls should be placed in the neutral position until power is restored.
9. The drill step should be resting on the bottom of the hole or on the platform with the stem secured to the mast before attempts are made to straighten a crossed cable on a reel.
10. While in operation, drills should be attended at all times.
11. Drill holes large enough to constitute a hazard should be covered or guarded.
12. Workers should not hold the drill steel while collaring holes, or rest their hands on the chuck or the centralizer while drilling.
13. Do not allow drilling crews to pressurize water tanks with air to force the

water through hoses or pipes to the drilling location for dust control, as water tanks are not designed to be pressure vessels.

## SAFETY PLANNING FOR BLASTING

Blasting during site preparation must be done under the following guidelines:[2,3]

1. The transporting, handling, storage, and use of dynamite and other explosive materials should be directed and supervised by a person of proven experience and ability in blasting operations. The operations should be in accordance with the requirements of manufacturers' recommendations and Department of Defense explosives safety standards where applicable.
2. Permission in writing should be initially obtained from the designated authority before explosive materials are brought on the job. Periodic replenishment of approved supplies does not require written approval.
3. Blasts using electrical detonators should be fired with an electric blasting machine or a properly designated power source. Blasts using nonelectric detonators should be fired by a blasting machine or starting device prescribed by the manufacturer. For blasting near radar or radio transmission facilities or near electrical energy sources where testing has shown that radio frequency (RF) energy or stray electrical current may present a hazard to electrical blasting, an approved nonelectrical initiation system should be employed. When electrical detonators are used, leg wires should be short-circuited (shunted) until connected into the circuit for firing.
4. Detonating cord should be initiated by nonelectric detonators (cap and fuse), electric detonators, shock-tube detonators, or gas-initiated detonators in accordance with the manufacturer's recommendations.
5. Delay electric detonators, nonelectric delay detonators, detonating cord connectors, or sequential blasting machines should be used for all delayed blasts. The practice should conform to the manufacturer's recommendations.
6. Operations involving the handling or the use of explosive materials should be discontinued and personnel moved to a safe area during the approach or the progress of a thunderstorm or a severe dust storm. Controls will be established to prevent accidental discharge of electric blasting caps from extraneous electricity.

7. Blasting machines should be maintained and used only as prescribed by the manufacturer.
8. Blasting machines should be tested prior to use and periodically thereafter as prescribed by the manufacturer.
9. Blasting machines should be secured and accessible only to the blaster. Only the blaster should connect the leading wire to the machine.
10. Warning signs should be provided at points of access to the blasting area.
11. When energy for blasting is taken from power circuits, the voltage should not exceed 550 volts. The wiring-control arrangements should conform to the following guidelines:
    a. The blasting switch should be an ungrounded UL-listed, enclosed, externally operated doublepole double throw switch that, when locked in the open position, will shunt the firing lines.
    b. A grounded switch should be installed between the blasting switch and the power circuit at a distance of not less than fifteen feet from the blasting switch.
    c. A lightning gap of at least fifteen feet should be provided between the two switches. The gap connection should be made by cable, plug, and receptacle.
12. The cable between switches should be disconnected, and both switches should be locked in the open position immediately after the shot is fired.
13. Keys to the switches should remain in the possession of the blaster at all times.
14. Insulated solid core wires of appropriate gauge in good condition should be used for all lines.
15. Sufficient firing line should be provided to permit the blaster to be located at a safe distance from the blast.
16. Blasting operations near overhead power lines, communications lines, utility services, or other structures should not be carried on until the operators and/or owners have been notified, and measures for safe control have been taken.
17. The use of black powder should be prohibited.
18. All loading and firing should be directed and supervised by one designated person.
19. A positive system to detect and measure the probability of lightning or massive static electrical discharges should be used.
20. No explosive materials should be abandoned.
21. Before any system of electrical firing is adopted, a thorough survey should be made for extraneous currents, and all dangerous currents should be eliminated before any holes are loaded.

22. All refuse from explosive loading such as empty boxes, paper, and fiber packing should not be reused for any purpose, but should be destroyed by burial at an approved location.

23. Mechanized equipment (including drills) should not be operated within fifty feet of a loaded hole.

24. Electromagnetic radiation creates specific hazards during blasting. The following requirements should be initiated:

    a. Blasting operations or storage of electrical detonators should be prohibited in the vicinity of operating RF transmitters or other RF-producing devices except where the clearances in ANSI C95.4 can be maintained.

    b. When it is necessary to perform blasting operations at distances less than those shown in ANSI C95.4 tables, an approved nonelectric initiation system should be used.

    c. Mobile radio transmitters, which are less than 100 feet away from electric blasting caps in other than original containers, should be deenergized and effectively locked, except in blasting areas where a nonelectric initiation system is used.

25. During blasting, vibration and damage control requires the following measures:

    a. Blasting operations in or adjacent to cofferdams, piers, underwater structures, buildings, structures, or other facilities should be carefully planned, with full consideration of all forces and conditions involved.

    b. Prior to the initiation of vibration-controlled blasting operations, a written plan for monitoring the operations should be established.

    c. When it is appropriate, owners, occupants, and the public should be notified of the nature of operations to be undertaken and controls to be established.

    d. Where vibration damage may occur, the energy ratios and peak particle velocities should be limited in accordance with state requirements. When any recording indicates that these criteria have been exceeded, blasting should be suspended, and the designated authority should be notified immediately. Blasting should not be resumed until the probable cause has been determined, and corrective measures have been taken.

    e. The air blast pressure exerted on structures resulting from blasting should not exceed 133 db (0.013 psi).

    f. The monitoring, recording, and interpreting of vibrations should be done by qualified personnel and should be consistent with state requirements.

    g. Records and interpretations should be furnished to the designated authority.

26. Transportation of explosive materials should be done as follows:
  a. Vehicles used for transportation of explosive materials should not be loaded beyond the rated capacity, and the explosive materials should be secured to prevent shifting of the load or dislodgment from the vehicle. When explosive materials are transported by a vehicle with an open body, a magazine or a closed container should be securely mounted on the bed to contain the cargo.
  b. All vehicles transporting explosive materials should display all placards, lettering, or numbering required by the U.S. Department of Transportation.
  c. Explosive materials and blasting supplies should not be transported with other materials or cargoes. Blasting caps (including electric) should not be transported in the vehicle or the conveyance with other explosives unless the conditions of 49 CFR 177.835(g) can be met.
  d. All vehicles used for transportation of explosive materials should be in the charge of and operated by a person who is physically fit, careful, reliable, able to read and understand instructions, and not under the influence of intoxicants or narcotics.
  e. No spark-producing tools, carbides, oils, matches, firearms, electric storage batteries, flammable substances, acids, or oxidizing or corrosive compounds should be carried in the bed or the body of any vehicle transporting explosive materials.
  f. Vehicles used in the transportation of explosives should be in good repair. When steel or part-steel bodies are used, nonsparking cushioning materials should separate the containers of explosives from the metal.
  g. Vehicles transporting explosive materials should be equipped with one or more fire extinguishers having a rating of 10-B:C and placed at strategic points. The extinguishers should be of a type listed by a nationally recognized test laboratory such as Underwriters' Laboratories, Inc. or Factory Mutual Engineering Corporation and should be ready for use. The driver will be trained in the use of the extinguisher.
  h. A vehicle containing explosive materials should not be taken into a garage or a repair shop, parked in a congested area, or stored at any time in a public garage or a similar building.
  i. All vehicles should be inspected before transporting explosive materials and all electric wiring completely protected and securely fastened to prevent short circuits. A written record of such inspection should be kept on file (DD Form 626).
  j. Vehicles transporting explosive materials should be operated with

extreme care. Full stops should be made at approaches to all railroad crossings and main highways, and the vehicles should not proceed until it is known that the way is clear.

k. Only the authorized driver and helper should be permitted to ride on any conveyance transporting explosive materials or detonators.

l. No vehicle should be refueled while explosive materials are in the motor vehicle except in an emergency.

m. Persons employed in the transportation, handling, or other use of explosive materials should not smoke or carry on their persons or in the vehicle matches, firearms, ammunition, or flame-producing devices.

n. Marine transportation of explosive materials should comply with the requirements of the U.S. Coast Guard.

o. Provision should be made for the safe transfer of explosive materials to magazine vessels, including substantial ramps or walkways free of tripping hazards.

p. Any explosive material transported by aircraft should be in the original unopened container.

q. Transportation of explosive materials over public highways should meet the provisions of Department of Transportation Regulations in 14 CFR, Part 103, "Air Transportation"; 46 CFR, Parts 146–149, "Water Carriers"; 49 CFR, Parts 171–179, "Highways and Railways"; and 49 CFR, Parts 390–397, "Motor Carriers."

r. Vehicles transporting explosive materials should not be left unattended.

s. The hoist operator should be notified before explosive materials are transported in a shaft conveyance.

t. Explosive materials should be hoisted, lowered, or conveyed in a powder car. No other materials, supplies, or equipment should be transported in the same conveyance at the same time.

u. No person should ride in any shaft conveyance transporting explosive materials. Loading and unloading should be accomplished only when the conveyance is stationary.

v. No explosive materials should be transported on any locomotive. At least two car lengths should separate the locomotive from the powder.

w. No explosive materials should be transported on a man haul trip.

x. The car or the conveyance containing explosive materials should be pulled, not pushed, whenever possible.

y. The powder car or the conveyance built for transporting explosive materials should bear on each side a reflectorized sign with a sharply contrasting background and the word "EXPLOSIVES" printed in letters not less than four inches high.

27. Handling of explosive materials needs to be done as follows:
    a. Containers of explosive materials should be opened only with non-sparking tools or instruments. Metal slitters may be used for opening fiberboard boxes, paper bags, or plastic tubes.
    b. Explosive materials should be removed from containers only as they are needed for immediate use.
    c. Explosive materials and detonators or primers should be separated and taken to the blasting area in original containers, Type 3 Magazines, or containers prescribed by 49 CFR 177.835.
    d. Primers should not be made up in excess of the immediate need for holes to be loaded.
    e. Primers should not be made up in or near magazines or excessive quantities of explosive materials.
    f. After loading of a blast is completed, all excess explosive materials and detonators should be removed to a safe location or returned at once to the storage magazines, observing the same rules as when being conveyed to the blasting area.
    g. The quantity of explosive materials taken to an underground loading area should not exceed the amount estimated to be necessary for the blast.
    h. Detonators and explosive materials should be taken separately into pressure working chambers.
28. Storage of explosives needs to be done as follows:
    a. Explosive materials and related materials should be stored in facilities required under the Alcohol, Tobacco, and Firearms Regulations contained in 27 CFR 55, "Commerce in Explosives."
    b. Magazines should meet the standards of the Bureau of Alcohol, Tobacco and Firearms as outlined in 27 CFR 55, Subpart K.
    c. Magazines used for storing detonators should be separated from magazines storing other explosive materials by the minimum distance for separation of magazines as specified. Detonators, primers, or other initiators should not be stored in the same magazine with explosives or blasting agents.
    d. The quantity of explosive materials permitted aboard a drill boat at any one time should not exceed the quantity required for one day's operation.
    e. The area around the magazine for a distance of twenty-five feet should be kept clear of vegetation and all combustible matter.
    f. All explosive materials should be accounted for at all times. Explosive materials not being used should be kept in a locked magazine, unavail-

able to persons not authorized to handle them. The employer should maintain an inventory and use record of all explosives. Authorities should be notified of any loss, theft, or unauthorized entry into a magazine.

g. Magazines in which explosive materials are stored should not be used for any other purpose.

h. No spark-producing agent of any description should be kept in magazines.

i. Explosive materials should be arranged in the magazine so that the oldest stock is used first.

j. Repairs to a magazine should not be made without first removing all explosive materials to a safe distance from it and providing them with protection.

k. Explosive materials should be stored only in their original containers.

l. Containers of explosive materials should be stored with the top side up so that the cartridges are lying flat and not standing on end.

m. Magazines should be kept dry and clean and the area posted with "EXPLOSIVES" signs unless the use of these signs is prohibited by a local authority. The signs should be located so that a bullet passing through a sign will not strike the magazine.

n. Explosive material containers, other than fiberboard, should not be opened, packed, or repacked inside or within fifty feet of a magazine.

o. Smoking, matches, firearms, open flames, or any flame-producing device should not be permitted within 100 feet of any magazine.

p. Magazine floors that become stained with nitroglycerin should be scrubbed with a nonsparking broom or brush using a freshly made solution composed of one-half gallon of water, one-half gallon of wood alcohol, and two pounds of sulfide of sodium. Sufficient liquid should be used to thoroughly decompose the nitroglycerin. Fire and explosion precautions must be taken.

q. Explosive materials should be located in conformance with the tables in 27 CFR 55, "Commerce in Explosives." Explosives on military sites should be stored in accordance with the DOD standard DOD 6055.9-STD.

r. Permanent underground storage magazines should be at least 300 feet from any shaft, adit, or active, underground working area.

29. Loading of explosives must meet these requirements:

a. The loading or loaded area should be kept free of any equipment, operations, or persons not essential to loading.

b. Tamping should be done with a wooden stick or an approved loading tool.

c. Primers should not be tamped.

d. Cartridges should be seated by even steady pressure only.

e. All loaded holes or charges should be checked and located before firing.

f. Loading of a sprung or jet-pierced hole should be prohibited until it is established that the hole has cooled sufficiently to allow loading.

g. Cartridges should be primed only in the number required for a single round of blasting.

h. No detonator should be inserted in explosive materials that do not have a cap well until a hole is made in the cartridge with a nonsparking punch of the proper size or with the appropriate pointed handle of a cap crimper.

i. All charges should be covered with blasting mats or back-covered before firing where blasting may cause injury or damage by flying rock or debris.

j. Where mats are used, care should be taken to protect electric blasting circuits.

k. In using a detonating cord downline, after the primer is loaded in the hole, the detonating cord should be cut from the supply reel before the rest of the charge is loaded.

l. Loaded holes should be stemmed to the collar with noncombustible material.

m. No explosive materials or loaded holes should be left unattended at the blast site.

n. A borehole should never be sprung when it is adjacent to or near a hole that is loaded. Flashlight batteries should never be used as a power source to replace a blasting machine when boreholes are sprung.

o. No holes should be loaded except those to be fired in the next round of blasting. After loading, all remaining explosive materials and detonators should be immediately returned to an authorized magazine.

p. Detonating cord should be handled and used with care to avoid damaging or severing the cord during and after loading and hooking up.

q. Detonating cord connections should be positive in accordance with recommended methods. Knot or other cord-to-cord connections should be made only with detonating cord in which the explosive core is dry.

r. All detonating cord trunklines and branchlines should be free of loops, sharp kinks, or angles that direct the cord back toward the oncoming line of detonation.

s. All detonating cord connections should be inspected before firing of the blast.

t. When detonating cord millisecond-delay connectors or short-interval-delay electric blasting caps are used with detonating cord, the practice should conform with the manufacturer's recommendations.

u. When a blasting cap is connected to detonating cord, the cap should be taped or otherwise attached securely along the side or the end of the detonating cord, with the charged end pointing in the direction in which the detonating cord will detonate.

v. Detonators for firing the trunkline should not be brought to the loading area or attached to the detonating cord until everything else is ready for the blast.

30. The wiring of explosives for blasting needs to meet these requirements:

a. The manufacturer's shunt should not be removed from the cap leg wires until the cap is connected to the lead line or to another cap in preparation for the assembly of two or more caps into a series circuit or when the cap is to be tested. When a single series of caps will be fired, or a number of series of caps will be fired as a series-in-parallel circuit, the resistance of the circuit should be checked with an approved blasting instrument (galvanometer, ohmmeter, or multimeter). For series-in-parallel circuits, each series should be "balanced," that is, have the same resistance. Each series circuit should be separately tested for two readings: (1) to ensure that the series is complete, and (2) to ensure that each series shows the same resistance and that this resistance is as close to the calculated resistance for such a series of caps as the testing instrument will read. If the first reading shows a series to be incomplete, the faulty cap or connection should be located and corrected. If the second reading shows an incorrect resistance, that is, higher or lower than expected, the cause should be found and corrected.

b. All caps in a blast should be of the same manufacture.

c. The firing line should not be smaller than a No. 14 B and S gauge solid copper conductor with a tough waterproof insulation.

d. The number of electric blasting caps in a circuit should not exceed the capacity of the blasting machine or the power source.

e. The circuit, including all caps, should be tested with an approved blasting instrument (galvanometer, ohmmeter, or multimeter) before being connected to a firing line.

f. No lead wire should be connected to the circuit until it has been grounded to dissipate any static charge.

g. The firing line should be checked with an approved blasting instrument (galvanometer, ohmmeter, or multimeter) before being connected to the blasting machine or other power source.

    h. No firing line should be connected to a blasting machine or other power source until the shot is to be fired.

31. The firing of explosives needs to meet these requirements:
    a. Prior to the firing of a shot, all persons in the danger area should be warned of the blast and ordered to a safe distance.
    b. The blast should not be fired until it is certain that every person has retreated to a safe distance and no one remains in a dangerous location.
    c. All blasting operations should use the following signals:
        (1) Warning signal: a one-minute series of long blasts five minutes prior to the blast signal.
        (2) Blast signal: a series of short blasts one minute prior to the shot.
        (3) All clear signal: a prolonged blast following inspection of blast area. The boat whistle on a drill boat should not be used as a blasting signal.
    d. The code of blasting signals and marking signs and flags should be posted at all access points.
    e. Prior to each shot, a competent flagperson should be posted at all access points to danger areas.
    f. Prior to firing, the drill boat or the vessel on the work should be moved a safe distance from the blasting range or area.
    g. Both before and while the drill boat or the vessel is being moved from the blasting area, a series of short signals by horn or whistle similar to the usual navigation warning signals should be given.
    h. No blast should be fired while any vessel under way is closer than 1,500 feet to the underwater blasting area. Those on board vessels or craft moored or anchored within 1,500 feet must be notified of firing before a blast is fired.
    i. No blast should be fired closer than 250 feet to a boat or a vessel containing an explosive magazine. Personnel engaged in drilling operations on another drill boat within 500 feet should leave the drill frames for cover if any holes have been loaded.
    j. No blast should be fired while any swimming or diving is in progress near the blasting area.
    k. Whenever a drill boat is moved from the drilling setting, all loaded underwater holes should be fired.
    l. The person making leading wire connections should fire the shot. All connections should be made from the borehole back to the source of the firing current, and the leading wire should remain shorted and not be connected to the blasting machine or another source of current until the charge is to be fired.

m. After the firing of an electric blast, the leading wires should be immediately disconnected from the firing power source and shunted.

n. In firing a circuit of electric blasting caps, care must be exercised to ensure that an adequate quantity of delivered current is available in accordance with the manufacturer's recommendations.

32. Inspection after blasting needs to meet the following requirements:

a. Immediately after the blast has been fired, the firing line should be disconnected from the blasting machine or the power source. Power switches should be locked open.

b. An inspection should be made by the blaster to determine that all charges have been exploded. All wires should be traced and a search made for unexploded cartridges.

c. Other persons should not be allowed to return to the area of the blast until an "all clear" signal is given.

d. Loose pieces of rock and other debris should be scaled down from the side of the face of excavation and the area made safe before proceeding with the work.

33. In the event of misfires, these requirements need to be met:

a. Misfires should be handled under the direction of the person in charge of the blasting. If a misfire is found, the blaster should provide proper safeguards for excluding all employees except those necessary to do the work from the danger zone.

b. No other work should be done except that necessary to remove the hazard of the misfire, and only those employees necessary to do the work should remain in the danger zone.

c. If broken wires, faulty connections, or short circuits are determined as the cause of a misfire, the repairs should be made, the firing line reconnected and tested, and the charge fired. This should be done, however, only after an inspection has been made of burdens remaining on such holes, and no hole should be so refired when the burden has been dangerously weakened by other shots.

d. No attempt should be made to extract explosive materials from any charged or misfired hole. A new primer should be put in and the hole refired. If refiring of the misfired hole represents a hazard, the explosive materials may be removed by washing out with water, or, where the misfire is under water, blown out with air.

e. Misfired charges tamped with solid material should be detonated by the following method where practical: Float out the stemming by the use of water, place a new primer, and detonate.

f. No drilling, digging, or picking should be permitted until all missed holes have been detonated, or the blaster has given approval for work to proceed.

34. Underwater blasting needs to meet these requirements:
    a. A blaster should conduct all blasting operations, and no shot should be fired without his/her approval.
    b. Loading tubes and casings of dissimilar metals should not be used, because of possible electric transient currents from galvanic action of the metals and water.
    c. Only water-resistant blasting caps and detonating cords should be used for all marine blasting. Loading should be done through a nonsparking metal loading tube when necessary.
    d. Blasting flags should be displayed.
    e. The storage and the handling of explosive materials aboard vessels used in underwater blasting operations should be done according to provisions contained in item 28 ("Storage of explosives") above.
    f. When more than one hole is loaded to be fired under water, a steel shotline should be anchored and floated over the row of loaded holes. The detonation downline from each loaded hole should be tied to the steel line and the loose end should be tied to the detonation trunkline. After the trunkline fires, the steel shotline should be inspected for misfires. Misfires should be handled in accordance with the requirements for misfires above (item 33).
    g. When there is drilling near or adjacent to a loaded hole, the drilling should be limited to vertical holes only and should be separated from loaded holes by the depth of water plus the depth of the loaded hole. If a solid casing or drill mast, vertically plumbed with an inclinometer, is extended from the barge and firmly seated on bedrock, the distance between a loaded hole and one being drilled should be one-third of the depth of the hole with a minimum distance of eight feet between the loaded hole and the one being drilled. Drilling should be halted to check the alignment with an inclinometer every four feet of the hole depth.

35. Blasting in excavation work under compressed air needs to meet these requirements:
    a. Detonators and explosive materials should not be stored or kept in tunnels, shafts, or caissons. Detonators and explosive materials for each round should be taken directly from the magazines to the blasting zone and immediately loaded. Detonators and explosive materials left over after the loading of a round should be removed from the working chamber before connecting wires are connected.
    b. When detonators or explosive materials are brought into an air lock, no

employee except the powderman, blaster, lock tender, and the employees necessary for carrying should be permitted to enter the air lock. No other material, supplies, or equipment should be locked through with the explosive materials.

c. Detonators and explosives should be taken separately into pressure working chambers.

d. The blaster or powderman should be responsible for the receipt, unloading, storage, and on-site transportation of explosive materials and detonators.

e. All metal pipes, rails, air locks, and steel tunnel lining should be electrically bonded and grounded at or near the portal or the shaft, and such pipes and rails should be cross-bonded at not less than 1,000-foot intervals throughout the length of the tunnel. In addition, each low air supply pipe should be grounded at its delivery end.

f. The explosive suitable for use in wet holes should be water-resistant and should be Fume Class 1.

g. When a tunnel excavation in rock face is approaching mixed face, and when the tunnel excavation is in mixed face, blasting should be performed with light charges and with a light burden on each hole. Advance drilling should be performed as the tunnel excavation in rock face approaches mixed face to determine the nature and the extent of the rock cover and the remaining distance ahead to soft ground.

## SAFETY PLANNING FOR ROADS

In the preparation of a site, roads need to meet the following requirements:

1. No contractor should move or cause to be moved construction equipment or vehicles upon any access, project, or haul roadway or grade unless the surface is constructed and maintained to accommodate safely the movement of the equipment and the vehicles involved. The contractor should construct necessary access and haul roads. A layout showing proposed road widths, travel directions, road markings, signposts, curves, grades, and sign distances should be submitted to the designated authority. At crossings, where grade separation is impractical, a plan of traffic control should be included.

2. When road levels are built on a hillside or next to water, berms, barricades,

or curbs should be constructed to prevent vehicles from overrunning the edge or the end of the embankment.

3. Roadways should have a crown and ditches for drainage.
4. Haul roads should be constructed to widths suitable for the safe operation of the equipment at the travel speeds proposed by the contractor and approved by the designated authority. All curves should have an open sight line and as great a radius as practicable. The vehicle speed should be limited on curves so that vehicles can be stopped within one-half of the visible distance of the roadway. The maximum allowable grade should be 10 percent except where permission to exceed this is granted by the construction manager.
5. Necessary lights (including traffic control lights), flagpersons, signs, barricades, and markings for the safe movement of traffic should be provided in accordance with the *Manual on Uniform Traffic Control Devices.*[4]
6. Dust controls should ensure safe operation at all times.

## REFERENCES

1. MacCollum, David V., *Crane Hazards and Their Prevention,* American Society of Safety Engineers, Des Plaines, IL, 1993.
2. E. I. du Pont de Nemours & Co., Wilmington, DE 19898, *Blasters' Handbook,* published yearly.
3. Hemphill, .Gary, *Blasting Operations,* McGraw-Hill Book Company, New York, 1981.
4. American National Standards Institute, *Manual on Uniform Traffic Control Devices for Streets and Highways,* D6.1, U.S. Government Printing Office, Washington, DC.

Chapter *23*

## Phase Planning for Excavating

If a man takes the cover off a pit or if he digs one and does not cover it, and a bull or a donkey falls into it, he must pay for the animal.

(Exodus 21:33–34)

Regardless of how small or how large the excavation is, or how little or how much reshaping of the terrain needs to be done before construction begins, planning is necessary to avoid cave-ins, injuries, and property damage. Phase planning must consider the type of excavation and the equipment to be used. The real bottom line of phase planning, regardless of the size of the project—be it construction of a small trench to bury pipes or utilities or the reshaping of the land to make footings for buildings, dams, tunnels, or harbors—is what is done on the job to prevent hazardous circumstances from arising and not how much paperwork is generated. The detailed recommendations listed in this chapter are intended to acquaint construction managers and staff with what must be considered in planning and what is pertinent to their projects so that injury, death, and property damage can be controlled. The most important function of planning is to ensure that those who need to know will know what has to be done and are held accountable for doing it. When planning assures that appropriate safeguards are in place, injury, death, and property damage are practically nonexistent. If safety planning is not done, how does a construction manager anticipate what safety measures are needed? Even worse, in the event of a disaster, if there has been no safety planning, there are no valid excuses for it.

Many of the following guidelines are based on the Corps of Engineers' *Safety and Health Requirements Manual,* EM 385-1-1.

## SAFETY PLANNING FOR ALL EXCAVATIONS

1. The sides of all excavations in which employees are exposed to danger from moving ground should be guarded by a shoring system, sloping of the ground, or other equivalent means. The minimum slope in any soil, with the exception of solid rock, should be no less than 3/4 horizontal to 1 vertical.
2. The determination of the angle of repose and the design of the supporting system should be based on careful evaluation of pertinent factors such as: depth of cut; possible variation in water content of the material while the excavation is open; anticipated changes in materials from exposure to air, sun, water, or freezing; loading imposed by structures, equipment, overlying material, or stored material; and vibration from equipment, blasting, or traffic.
3. Where a weak stratum is overlain by strata with a greater angle of repose than its own, the angle of repose for the overlying strata should be taken as no more than that of the supporting stratum.
4. The design of the supporting system (i.e., piling, cribbing, shoring, etc.) should be based on the calculation of forces and their direction, with adequate consideration given to surcharges, the angle of internal friction of materials in the face, and other pertinent properties of the material to be retained in place. When tight sheeting or sheet piling is used, full loading due to the groundwater table should be assumed unless prevented by weep holes, drains, or other means. Additional stringers, ties, and bracing should be provided to allow for any necessary temporary removal of individual supports.
5. Except in hard rock, excavations below the level of the base of the footing of any foundation or retaining wall should not be permitted unless the wall is underpinned and all other necessary precautions are taken to ensure the stability of the adjacent walls, for the protection of the employees involved in the excavation work or in the vicinity thereof. If the stability of adjoining buildings or walls is endangered by excavations, then shoring, bracing, or underpinning designed by a qualified person should be provided. Such

shoring, bracing, or underpinning should be inspected daily or more often, as conditions warrant, by the qualified person, and the protection should be effectively maintained.

6. Division ditches, dikes, or other means should be used to prevent surface water from entering an excavation and to provide good drainage of the area adjacent to the excavation.

7. Excavated material should be stored and retained at least two feet from the edge of the excavation and at a distance to prevent excessive loading on the face of the excavation.

8. Boulders, stumps, or other materials that may slide or roll into the excavation should be removed or made safe.

9. Guardrails, fences, or barricades and warning lights or other illumination, maintained from sunset to sunup, should be placed at all excavations that are adjacent to paths, walkways, sidewalks, driveways, and other pedestrian or vehicle thoroughfares. Adequate physical protection should be provided at all remotely located excavations. All wells, calyx holes, pits, shafts, and so on, should be barricaded or covered. Upon completion of exploration and similar operations, temporary wells, calyx holes, pits, shafts, and so on, should be backfilled immediately.

10. Walkways or bridges with guardrails should be provided where people or equipment will be required or permitted to cross over excavations.

11. Materials used for sheeting, sheet piling, cribbing, bracing, shoring, and underpinning should be in good serviceable condition and of adequate dimensions. Timbers should be sound and free of large or loose knots.

12. Side slopes and faces of all excavations should be maintained in safe condition by scaling, ice removal, benching, barricading, rock bolting, wire mesh, or other means. Special attention should be given to slopes that may be adversely affected by weather, moisture content, or vibration.

13. Precautions should be taken in sloping or shoring the sides of excavations adjacent to a previously backfilled excavation or a fill, particularly when the separation is less than the depth of the excavation. Attention also should be paid to joints and seams of material comprising a face and the slope of such seams and joints.

14. Support systems should be planned and designed by a qualified person when an excavation is more than twenty feet deep, adjacent to structures or improvements, or subject to vibration or groundwater.

15. Prior to the opening of an excavation, underground installations (i.e., sewer, water, fuel, electric lines, etc.) should be located and protected from damage or displacement. Utility companies should be contacted to have them locate and mark actual locations.

16. Where unexploded munitions/military ordnance or hazardous materials

might be uncovered during excavation, job planning will require contact with local safety/health authorities.

17. Where personnel are required to enter excavations over four feet deep, sufficient stairs, ladders, or ramps should be provided to require no more than twenty-five feet of lateral travel. When access to excavations more than twenty feet deep is required, ramps, stairs, or mechanical personnel hoists should be provided.

18. Where it is necessary to undercut the side of an excavation, overhanging material should be safely supported.

19. Bracing, shoring, cribbing, and other supports should be inspected daily and after every rainstorm or other hazard-increasing occurrence by a competent person. Protection against slides and cave-ins should be increased if necessary. If evidence of possible cave-ins or slides is apparent, all work in the excavation should cease until the necessary precautions have been taken to safeguard the employees.

20. When tie-rods are used to restrain the top of sheeting or another retaining system, the rods should be anchored well back of the angle of repose.

21. In locations where oxygen deficiency or gaseous conditions are known or suspected, air in the excavation should be tested prior to the start of each shift or more often if so directed by the designated authority. A log of all test results should be maintained at the work site.

22. Emergency rescue equipment such as a breathing apparatus, safety harness and line, and basket stretcher should be readily available where adverse atmospheric conditions are suspected or may develop in an excavation.

23. Where ramps are used, a minimum width of four feet for personnel and twelve feet for equipment should be provided. Guardrails should be provided on all personnel ramps. Curbs of not less than 8″ × 8″ timbers or equivalent protection should be provided on equipment ramps. Such ramps should be designed and constructed in accordance with accepted engineering practice.

24. When mobile equipment is utilized or allowed adjacent to excavations, substantial stop-logs or barricades should be installed.

25. Excavating or hoisting equipment should not be allowed to raise, lower, or swing loads over personnel in the excavation without substantial overhead protection.

26. At least two means of exit should be provided for personnel working in excavations. Where the width of the excavation exceeds 100 feet, two or more means of exit should be provided on each side of the excavation.

27. Groundwater should be controlled. Freezing, pumping, drainage, and similar control measures should be planned and directed by a competent engineer. Consideration should be given to the existing moisture balance

in surrounding soils and the effects on foundations and structures if it is disturbed. When continuous operation of groundwater control equipment is necessary, an emergency power source should be provided.

## SAFETY PLANNING FOR TRENCH EXCAVATION

1. Excavation of trenches under powerlines may be a very dangerous activity. See Chapter 22, in the section on powerlines for some general recommendations.
2. Banks more than five feet high should be shored, laid back to a stable slope, or provided with other equivalent protection where employees may be exposed to moving ground or cave-ins. Trenches less than five feet deep also should be protected when examination of the ground indicates that hazardous ground movement may be expected. The safe angle of repose for soil conditions and bracing systems should be determined by a qualified person.
3. Bracing or shoring of trenches should be carried out along with the excavation.
4. Cross braces or trench jacks should be in true horizontal position, secured to prevent sliding, falling, or kick-outs. (See Figure 23-1.)
5. Portable trench boxes, sliding trench boxes, or shields should be designed, constructed, and maintained in such a manner as to provide protection equal to or greater than that of the sheathing and shoring required for the situation.
6. Ladders used as accessways should extend from the bottom of the trench to not less than three feet above the surface. Lateral travel to an exit ladder should not exceed twenty-five feet.
7. Backfilling and removal of trench supports should progress together from the bottom of the trench. Jacks or braces should be released slowly, and in unstable soil ropes should be used to pull out the jacks and braces from above after personnel have cleared the trench.
8. The minimum size and spacing of timbers for shoring of trenches should be in accordance with minimum requirements for trench shoring.
9. Braces and diagonal shores in a wood shoring system should not be subjected to compressive stress in excess of values given by the following formula:

$$S = 1300 - \frac{20L}{D}$$

$$\text{Maximum ratio: } \frac{L}{D} = 50$$

where:
$L$ = Length of the timber, unsupported, in inches.
$D$ = Least side of the timber, in inches.
$S$ = Maximum allowable stress in pounds per square inch of cross section.

10. Aluminum hydraulic shoring should be installed in accordance with the manufacturer's recommendations and the requirements listed in the preceding section (on planning for all excavations).
11. OSHA has specific requirements for a competent person familiar with OSHA §1926.652 who:
    a. Knows how to do soil classifications.
    b. Knows about hazardous atmospheres.
    c. Has received the training required to pass an OSHA inspection.
    d. Is capable of identifying existing or predictable hazards in the excavation's surroundings.
    e. Recognizes working conditions that are unsanitary or otherwise precarious to workers.
    f. Will conduct a preconstruction site review to develop a strategy that ensures a safe, efficient job process.
    g. Performs inspections of equipment and trench conditions at the start of each shift or as needed by changing conditions.
    h. Removes employees from hazardous conditions and effects all changes necessary to ensure safety.
    i. Categorizes soil conditions and conducts visual and manual tests to determine stability of the soil and surrounding trench conditions.
    j. Maintains on-site records of protection systems.
    k. Examines all equipment to verify conditions.
    l. Determines the appropriate protective system and oversees its installation.
    m. Verifies that a competent person designs ramps and walkways for employee use in accordance with OSHA rules.
    n. Verifies that a competent person qualified in structural design builds equipment ramps.
    o. Holds toolbox sessions with the crew before trenching and shoring take place as required.
    p. Assures that emergency rescue equipment is available to meet existing or potential conditions.
    q. Monitors the use of water-removal equipment.

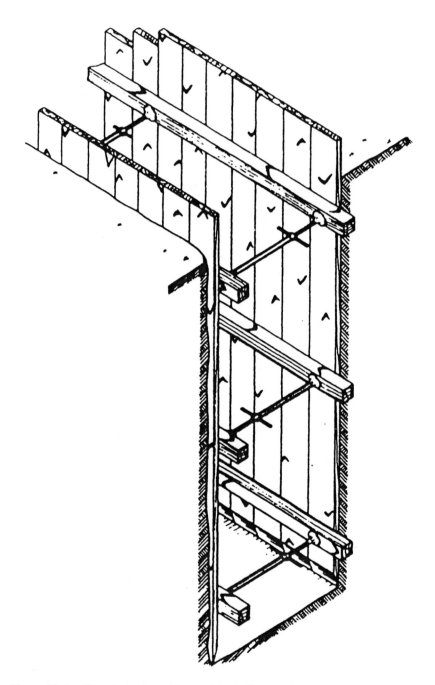

**Figure 23-1.**   Trench shoring with trench jacks for cross braces.

r. Tests for the presence of oxygen and air quality in excavations deeper than four feet as necessary.

s. Consults with a registered professional engineer for trenches more than twenty feet deep for specially designed shoring, bracing, or underpinning, or when excavation endangers nearby structures.

## SAFETY PLANNING FOR EXCAVATION
## OF SMALL-DIAMETER SHAFTS

1. Calyx holes, caisson footings, and other small-diameter shafts that persons are required to enter should be provided with a steel casing of sufficient strength to withstand shifting of the surrounding earth. The casing should be provided the full depth of the overburden and should extend at least five feet into solid rock where such conditions exist.

2. The steel casing should extend at least one foot above ground level and be equipped with a cover that can be locked. This cover should be in place whenever no work is being performed in the excavation and until the excavation is backfilled to ground level.

3. Before work in a shaft, the air at the bottom of the shaft should be tested for oxygen deficiency and harmful contaminants. No work should be done in the shaft until the air is safe for breathing.

4. Mechanical ventilation should be used to change the air in shafts. Compressed air from an untested source should not be used.

5. Hoisting of persons should be done only with equipment meeting applicable requirements.

6. An enclosed covered metal cage should be used to raise and lower persons in the shaft. The cage should be designed with a safety factor of four, and should be load-tested prior to use. The exterior of the cage should be free of projections or sharp corners. Only closed shackles should be used in the cage rigging.

7. If the cage is equipped with a door, a positive locking device should be installed to prevent the door from opening accidentally while the cage is being lowered or raised.

8. The rigging and the cage should be checked at the start of each shift by lowering the empty cage to the bottom of the shaft and hoisting it to the surface.

9. If the work or an emergency prohibits the use of a cage, persons may be raised and lowered in the shaft in the muck bucket or a boatswain's chair.

Persons hoisted in this manner should wear a parachute-type harness with an independent safety line attached. "Riding the hook" should not be permitted.

10. A tender should be stationed at the shaft entrance whenever personnel are in the shaft, and should not be assigned other duties.

11. A communication or signal system should be provided and used.

12. Procedures for rescuing personnel who might be injured or incapacitated while in the shaft should be written and posted at the job site. These procedures should include hoisting methods to be used in the event that the cage and the cage hoisting device are inoperative.

13. No employee should be permitted to enter an unsupported auger-type excavation in unstable material for any purpose. Necessary clean-out in such cases should be accomplished without entry.

14. All wells or shafts over five feet deep should be retained with lagging, piling, or casing.

15. In shafts, the employer should inspect the walls, ladders, timbers, blocking, and wedges of the last set to determine if they have loosened following blasting operations. If they are found unsafe, corrections should be made before shift operations are started.

16. Safety belts should be worn on skips and platforms used in shafts by crews when the skip or the cage does not occlude the opening to within one foot of the sides of the shaft, unless guardrails or cages are provided.

## Safety Planning for Tunneling

Tunneling has long been considered inherently dangerous work. Phase planning should include medical requirements and procedures for the safe erection of portals, working in soil, underground blasting, and haulage.[1,2]

## Compressed Air Tunneling and Caissons

1. Because compressed air work is exceedingly dangerous, as a minimum, the following medical requirements are necessary:
   a. Requirements for a medical lock and its operation.

    b. An identification system for compressed air workers.

    c. Communications system requirements.

    d. Requirements for signs and recordkeeping.

    e. Special compression and decompression requirements.

    f. Man lock and decompression chamber requirements.

    g. Requirements for compressor systems and air supply.

    h. Ventilation requirements.

    i. Electrical power requirements (explosion-proof equipment).

    j. Sanitation considerations.

    k. Fire prevention and fire protection considerations.

    l. Requirements for bulkheads and safety screens.

   m. Retaining one or more licensed physicians who are familiar with and experienced in the physical requirements and the medical aspects of compressed air work and the treatment of decompression illness. A physician should be available at all times while work is in progress in order to provide medical supervision of workers doing compressed air work. All physicians should be physically qualified and be willing to enter a pressurized environment.

    n. Provision of a fully equipped first aid station at each tunnel project regardless of the number of persons employed. An ambulance or transportation suitable for a litter case should be provided at each project.

    o. If compressed air (for breathing) is used, where tunnels are being excavated from portals more than five road miles apart, provision of a first aid station and transportation facilities at each portal.

2. Whenever, in caisson work in which compressed air is used, the working chamber is less than eleven feet long, and when such caissons are at any time suspended or hung while work is in progress so that the bottom of the excavation is more than nine feet below the deck of the working chamber, a shield should be erected therein for the protection of the employees.

3. Shafts should be subjected to a hydrostatic test, at which pressure they should be tight. The shaft should be stamped on the outside shell about twelve inches from each flange to show the safe working pressure.

4. Whenever a shaft is used, it should be provided, where space permits, with a safe, proper, and suitable staircase for its entire length, including landing platforms (not more than twenty feet apart). Where this is impractical, ladders not more than twenty feet high should be installed with each section offset from adjacent sections and a guarded landing provided at each offset.

5. All caissons having a diameter or side greater than ten feet should be provided with a man lock and shaft for the exclusive use of employees.

6.  In addition to the gauge in the locks, an accurate gauge should be maintained on the outer and the inner side of each bulkhead. These gauges should be accessible at all times and kept in accurate working order.

## SAFETY PLANNING FOR TUNNEL PORTAL PREPARATION

1.  Rock faces above and adjacent to portal areas should be kept thoroughly scaled and all loose or overhanging rocks removed.
2.  Rock masses separated from the main mass by faults, joints, or fractures should be secured by rock bolting or other suitable means or should be removed. The means of securing should be designed by a foundation engineer and an engineering geologist or other qualified person.
3.  An anchored chain-link fabric or another approved method should be provided on rock faces subject to spalling.
4.  Where tunnels are excavated through earth or shale, the excavations above and adjacent to portal areas should be sloped to the angle of repose or held in place by ground supports. When undercutting occurs on these slopes, due to erosion or other causes, the overhanging material should be removed promptly.
5.  Ice and snow buildup on rock faces or earth slopes that create a hazard should be removed promptly.
6.  Where the need is indicated, a protective shelter should be provided at each tunnel portal to protect persons and equipment from the hazards of falling rock or other material. The protective shelter should project at least fifteen feet from the tunnel portal.
7.  Torque meters and torque wrenches should be available in tunnels where rock bolts are used for ground support. Frequent tests should be made to determine if bolts meet the required torque. The test frequency should be determined by rock conditions and the distance from vibration sources. Rock bolt support systems should be designed by a foundation engineer, a geologist, or another qualified person.
8.  All sets, including horseshoe or arched rib steel sets, should be designed and installed so that the bottoms will have required anchorage to prevent pressure from pushing them inward into the excavation. Lateral bracing should be provided between sets to further stabilize the support.
9.  Damaged or dislodged tunnel supports, whether steel sets or timber, should be repaired or replaced. New supports should be installed whenever possible before removal of the damaged supports.

## GENERAL REQUIREMENTS FOR TUNNELING

1. Vehicles not directly involved in the work should be kept away from the portals and separated from construction activities.
2. Noise control and ear protection should be provided.
3. Personnel in wet underground areas should wear safety-toed rubber boots and rain gear.
4. Lighting circuits should be installed on one side of the tunnel near the spring line and should be mounted on insulators at each point of suspension.
5. Light fixtures should be nonmetallic and weatherproof and so mounted to provide safe clearance for persons and equipment.
6. Ventilation and exhaust systems for tunnel excavation should be of sufficient capacity to maintain an adequate supply of uncontaminated air at all points in the tunnel. The supply of fresh air should not be less than 200 cubic feet per minute for each employee underground plus that necessary to operate the equipment. The linear velocity of the air flow in the tunnel bore should not be less than 30 feet per minute in those tunnels where blasting or rock drilling is conducted or where there are other conditions that are likely to produce dusts, fumes, vapors, or gases in harmful quantities.
7. Atmospheric testing devices should be provided and used as applicable to determine the oxygen content of the air and concentrations of flammable or toxic gases that occur in the tunnel or the shaft. These tests should be made on a fixed schedule but not less than once each shift. If a 1.5 percent or higher concentration of flammable gas is detected in the air of underground workplaces, the employees should be withdrawn until the flammable gas concentration is reduced to 1 percent or less. (An MSHA-approved explosimeter may be used in lieu of the percent concentration meter.) These tests should be made by qualified personnel.
8. A log of all atmospheric tests should be maintained at the work site. The log will show the date and the time of all tests, test results, and the name of the person conducting the tests.
9. Storage of combustible or flammable material and the use of open fires in tunnels should be prohibited. Heating devices used in tunnels should be approved by the designated authority.
10. Walkways should be provided for personnel access to all underground work areas. Work areas and access walkways should be maintained free of debris, waste material, and stumbling hazards.
11. Nonslip decking should be provided on all working platforms of drill jumbos. Working platforms should be provided with a standard railing.

12. Safe access should be provided to all working levels of the drill jumbos.
13. All connections between sections of "bull hose," used to supply air to drill jumbos, should be provided with stay chains or lashing to prevent whipping should they become disconnected. Air lines that are buried in the invert should be identified by signs posted nearby, warning personnel.
14. A program for testing all rock bolts for tightness should be established. The frequency of testing should be determined by rock conditions and the distance from vibration sources.
15. The employer should examine and test the roof, face, and walls of the work area at the start of each shift and frequently thereafter.
16. Gasoline or liquefied petroleum gases should not be taken, stored, or used underground. Not more than one day's supply of diesel fuel may be stored underground.
17. Air that has passed through underground oil or fuel-storage areas should not be used to ventilate work areas.
18. Fire-resistant hydraulic fluids approved by a nationally recognized authority or agency such as Underwriters' Laboratories, Inc. or Factory Mutual Laboratories should be used in hydraulically actuated underground machinery and equipment. A properly designed fire protection system may be installed in lieu of fire-resistant hydraulic fluids.
19. Drilling equipment that is to be used during a shift should be inspected on each shift by a competent person. Equipment defects affecting safety should be corrected before the equipment is used.
20. Before the drill cycle starts, the face and the lifters should be examined for misfires (residual explosives), and, if found, they should be removed before drilling commences. Lifters should not be drilled through blasted rock (muck) or water.
21. Scaling bars should be in good condition at all times. Blunted or severely worn bars should not be used.
22. When jumbos are being moved, riders should not be allowed on the jumbo unless they are assisting the driver.
23. On jumbo decks over ten feet in height, removable guardrails (pipe in sockets with chain handrail), or their equal, should be provided on all sides and back platforms.
24. The employer should provide stair access to jumbo decks wide enough to accommodate two persons for decks over ten feet in height.
25. When a drill is being moved from one drilling area to another, the drill steel, tools, and other equipment should be secured and the mast placed in a safe position.
26. The drilling area should be inspected for hazards before the start of the drilling operation.

27. Employees should not be allowed on a drill mast while the drill bit is in operation.
28. Receptacles or racks should be provided for drill steel stored on jumbos.
29. Before drilling is started, a warning should be given to personnel working below jumbo decks.
30. Drills on columns should be anchored firmly before drilling is started and should be retightened frequently.
31. The employer should provide mechanical means for lifting drills, roof-bolts, mine strips, and other heavy material to the top decks of jumbos over ten feet in height.
32. Oil-filled transformers should not be used underground unless they are located in a fire-resistant enclosure and surrounded by a dike to contain the contents of the transformers in the event of a rupture.
33. Access to unattended underground openings should be restricted by gates or doors.
34. Each tunneling operation should have a check-in and check-out system that will provide a positive identification of every employee underground.
35. Emergency hoisting facilities should be readily available at shafts more than fifty feet deep, unless hoisting facilities are provided that are independent of electrical power failures.
36. U.S. Bureau of Mines approved self-rescuers should be available near the advancing face to equip each face employee. Such equipment should be on the haulage equipment and in other areas where employees might get trapped by smoke or gas, and should be maintained in good condition.
37. Telephone or other signal communication should be provided between the work face and the tunnel portal, and such systems should be independent of the tunnel power supply.
38. Ventilation doors, not operated mechanically, should be designed and installed so that they are self-closing and will remain closed regardless of the direction of the air movement.
39. When air flow is resumed in a short time after ventilation failure, all places where flammable gas may have accumulated should be examined by a competent person and determined to be free of flammable gas before power is restored and work resumed. If the ventilation has been shut down completely with all employees out of the shaft, only a person specified by the designated authority may go underground until after the fans have been restarted and the work areas declared safe.
40. Noncombustible barriers should be installed below welding or burning operations in or over a shaft or raise.

## SAFETY PLANNING FOR TUNNELING IN SOFT GROUND

1. Where tunnels are excavated by conventional methods, the excavation should not be extended more than twenty-four inches in advance of tunnel supports.
2. Where continuous mining machines are used for tunnel excavation, the excavation should not be extended more than forty-eight inches in advance of tunnel supports.
3. Under no circumstances should persons be permitted to work in unsupported sections of the tunnels.
4. All voids back of ring beams, liner plates, timbering, and other tunnel supports should be filled, blocked, braced, or treated to prevent further cave-ins.
5. Where liner plate is not used for tunnel support, two-inch wire mesh or chain-link fabric should be installed over the crown section, extending down to the spring line on each side of the tunnel. It should be held in place securely.

## SAFETY PLANNING FOR BLASTING IN HARD ROCK

In addition to the basic requirements for blasting defined in Chapter 22, special requirements for underground blasting in hard rock must be included in planning:

1. Dynamite used in tunnel blasting should be Fume Class 1. Fume Class 2 and Fume Class 3 explosives may be used if adequate ventilation is provided.
2. Prior to loading, all power, water, and air lines should be disconnected from the loading jumbo. Powerlines, including lighting circuits, should be moved back a minimum of fifty feet.
3. The loading area should be illuminated (minimum ten foot-candles) by floodlights located fifty feet from the face. If additional illumination is needed, the loading crew should be provided with head lamps approved by the U.S. Bureau of Mines.
4. Equipment used for pneumatic placement of non-cap-sensitive blasting agents should be designed for that purpose and should be grounded while in use.

5. Storage of explosives, blasting agents, and detonators in tunnels or underground work areas should be prohibited.
6. Blasting power circuits should be separate and distinct from other power and lighting circuits.
7. Sectioning switches or the equivalent should be installed in the firing line at 500-foot intervals.
8. No blasts should be fired until all persons have cleared the blast area and are in a safe position.
9. The person in charge of blasting should be the last to leave the blast area. That person should see that no one remains in the blast area and will operate the sectioning switches in the firing line while proceeding out of the blast area.
10. All underground blasts fired by external power should be fired by a power blasting switch system.
11. No persons should enter the tunnel blast area until the ventilation system has cleared the heading of harmful gases, smoke, and dust.
12. Immediately after each blast, the rock surfaces should be inspected, scaled, and, if required, provided with shoring, bracing, rock bolts, shotcrete, or chain-link fabric before mucking is started.
13. Rock bolts within 100 feet of a blast should be tested after each blast before drilling for the next round begins.
14. The muck pile should be wet down prior to mucking and kept wet during mucking operations.
15. Trucks used for the transportation of explosives underground should have the electrical system checked weekly to detect any failures that may constitute an electrical hazard. A written record of such inspections should be kept on file. The installation of auxiliary lights on truck beds, powered by the truck's electrical system, should be prohibited.
16. Explosives or blasting agents not in the original containers should be placed in a suitable container when transported manually. Detonators, primers, and other explosives should be carried in separate containers when transported manually.

## SAFETY PLANNING FOR UNDERGROUND HAULAGE

1. Internal combustion engines other than mobile diesel should not be used underground. Mobile diesel-powered equipment used underground should

be certified by MSHA U.S. Department of Labor, according to 30 CFR, Part 32, "Mobile Diesel-Powered Equipment for Noncoal Mines."

2. Rubber-tired haulage equipment should be equipped with two headlights at each end, a backup light, and an automatic backup alarm.

3. All vehicles or equipment required to move in and out of tunnels should have a revolving, flashing amber light, mounted so as to be visible in all directions. The flashing light will be on when the equipment is in operation.

4. Where narrow-gauge railroads are used for haulage, the tracks should be secured to prevent shifting. No "humping" of mine dump cars should be permitted.

5. Mine dump cars should be equipped with automatic safety couplings, and cradle cars should be equipped with a positive locking device to prevent accidental dumping.

6. Conveyors used to transport muck from tunnels should be installed, guarded, and maintained. Fire extinguishers or equivalent protection should be provided at the head and tail pulleys of underground belt conveyors and at 300-foot intervals along the belt line.

7. When cars are dumped by hand, the car dumps should be provided with tie-down chains or bumper blocks to prevent the cars from overturning.

## SAFETY PLANNING FOR PILEDRIVING

Some excavations require that piling be driven or cofferdams constructed. The following guidelines should be followed to ensure safe piledriving procedures:

1. The width of hulls of floating piledrivers should not be less than 45 percent of the height of the lead above the water. The operating deck of floating piledrivers should be so guarded as to prevent piles that are being hoisted into driving position from swinging in over the deck.

2. Dogs on piledriver hoist drums that automatically disengage by either relieving the load or rotating the drum should be prohibited.

3. All hose connections to piledriver hammers, pile ejectors, or jet pipes should be securely attached with an adequate length of at least ¼-inch chain, having a 3,250-pound working load limit, alloy steel or equal-strength cable, to prevent whipping if the joint is broken.

4. Hanging or swinging leads of piledrivers should have fixed ladders.

Employees should be prohibited from remaining on leads or ladders while a pile is being driven. Fixed leads should be provided with rings or attachment points so that an aloft worker may engage his/her safety belt lanyard to the leads.

5. Fixed piledriver leads should be provided with decked landings having guardrails, intermediate rails, and toeboards. Fixed ladders or stairs should be provided for access to landings and head blocks.

6. Landings or leads should not be used for storage of any kind.

7. Piledriver leads should have stop blocks to prevent the hammer from being raised against the head block.

8. Pilehammers should be lowered to the bottom of leads while the piledriver is being moved.

9. Guys, outriggers, thrustouts, counterbalances, or rail clamps should be provided to maintain the stability of piledriver rigs.

10. Taglines should be used for controlling "unguided piles" and "flying hammers."

11. Hoisting of steel piling should be done by use of a closed shackle or other positive attachment that will prevent accidental disengagement.

12. If piling cannot be pulled without exceeding the load rating of equipment, it should not be done.

13. When pulling piling, crane booms should not be raised more than 60 degrees above the horizontal.

14. Piling should not be pulled by tipping the crane, releasing the load brake momentarily, and catching the load before the crane has settled.

15. Overhead protection should be provided that will not obscure the vision of the operator, and it should be the equivalent of two-inch planking or other material of equivalent strength.

16. A blocking device, capable of supporting the weight of the hammer, should be provided for placement in the leads under the hammer at all times while employees are working under the hammer.

17. Guards should be provided across the top of the head block to prevent the cable from jumping out of the sheaves.

18. Steam-line controls should consist of two shutoff valves, one of which should be a quick-acting lever type within easy reach of the hammer operator.

19. When it is necessary to cut off the tops of driven piles, piledriving operations should be suspended except where the cutting operations are located at least twice the length of the longest pile from the driver.

20. All pressure vessels that are a part of, or used with, piledriving equipment should meet the requirements of the American Society of Mechanical Engineers, *Pressure Vessels* (Section VIII).

21. All employees should be kept clear when piling is being hoisted into the leads.
22. In driving jacked piles, all access pits should be provided with ladders and bulkheaded curbs to prevent material from falling into the pit.
23. Vibratory piledrivers/extractors should not be used without the permission of the crane manufacturer. This equipment does not require leads; therefore, in driving or extracting piles, the load is not freely suspended as it is in normal crane operations, and the crane operator, especially during extraction, has no way of knowing if he/she is exceeding the rated capacity of the crane, which can result in either upset or buckling of the boom. Even if a load-moment device or a load indicator is used, the manufacturer still should be contacted to determine whether the vibratory equipment can be used in the particular situation.

## SAFETY PLANNING FOR EXCAVATION WITH COFFERDAMS

1. If overtopping of the cofferdams by high water is possible, the design should include provisions for controlled flooding of the work area.
2. Where personnel or equipment will be required or permitted on cofferdams, walkways, bridges, or ramps, standard railings or equivalent protection should be provided.
3. At least two ladders, walkways, ramps, roads, or other means of rapid exit should be provided for personnel and equipment working on cofferdams.
4. A plan including a warning signal for the excavation of personnel and equipment in case of emergency and for controlled flooding should be developed and posted.
5. Cofferdams located close to navigable shipping channels should be protected from vessels in transit, if possible.

## SAFETY PLANNING FOR DRAGLINE EXCAVATION

Dragline operations take their toll in injury, death, and property damage. Workers removing mud from the crawler tracks have been crushed in the narrow clearance between the rotating cab and the tracks when the cab has been unexpectedly rotated; operators have been injured or killed when draglines have upset from overloading because the bucket was too large and too much material was scooped up during casting; and operators have been drowned and draglines damaged when unstable or water-soaked ground has given way during excavation of materials from under water, and the dragline has tipped into the water.

To avoid injury, death, and property damage on a project, planning should include the following:

1. Draglines should be thoroughly examined upon arrival at a project, in accordance with guidelines found in Chapter 14.
2. Dragline work should never be allowed adjacent to powerlines (see Chapter 22).
3. Barge-mounted draglines should comply with U.S. Coast Guard requirements.
4. To avoid having someone caught in the narrow clearance between the bottom of the cab and the tracks when cleaning out mud, fence off the swing area to limit access while the dragline is working; require that cleaning of tracks for mud only be done after the dragline is shut down; or select only those draglines that have sufficient clearance between the cab and the tracks.
5. Be sure that the dragline bucket is not so large that the amount of material in it can overload and upset the dragline.
6. Position the dragline on ground that is stable and not water-soaked to avoid upset.

## SAFETY PLANNING FOR USE OF SCRAPERS

1. To assure they are in safe operating condition, scrapers brought onto a project should be examined as outlined in Chapter 14.
2. Be sure that scrapers are maintained in safe condition, as poorly maintained scrapers have been known to cause upset or loss of control.
3. Scrapers should be equipped with ROPS to avoid injury if they overturn on uneven or steep ground, and with a screen behind the operator to prevent the intrusion of clods of dirt, as discussed in Chapter 14.
4. Haul roads should not be built with excessive grades or roughness.
5. Provide state-of-the-art shock-absorbing seats with full back rests and functional restraint systems (seat belts and shoulder harnesses) to avoid operator injury due to whiplash in push–pull operations on rough haul roads.

## SAFETY PLANNING FOR USE OF BULLDOZERS, FRONT-END LOADERS, GRADERS, AND COMPACTORS

1. Examine these machines to see that they are in safe operating condition when first brought onto a project, as outlined in Chapter 14.
2. Bulldozers, front-end loaders, graders, and compactors should be equipped with ROPS to avoid injury if rollover should occur.
3. Regular, scheduled maintenance should be done to keep these machines in a safe condition.
4. When they are being used on haul or other roads, plan for other vehicular traffic to be rerouted or banned.
5. Where public traffic must be accommodated, follow the guidelines set forth in the *Manual on Uniform Traffic Control Devices for Streets and Highways,* ANSI D6.1.[3]

## SAFETY PLANNING FOR EXCAVATORS

An excavator operator was drowned when the excavator, which was positioned on unstable, water-soaked ground next to a pond, upset into the water, and the operator was imprisoned and could not escape because there were no emergency exits.

The following guidelines apply for excavator use:

1. All excavators and shovels brought onto a project should be examined as outlined in Chapter 14, and maintained in safe operating condition.
2. Be sure that excavators are positioned on stable, dry ground when working adjacent to water.
3. Be sure that excavators have emergency, knockout windows for emergency exit.

## SAFETY PLANNING FOR SHOVELS AND HAUL TRUCKS

1. When first brought onto a project, shovels and haul trucks should be examined to see that they are in safe operating condition.

2. Small pickup-truck traffic should not be allowed around shovels or haul trucks.
3. Radio communication should be established between operators of haul trucks, shovels, and any other vehicles in the work area.
4. Shift changes should be planned to conform to times when they will not interfere with loading operations to lessen pedestrian and vehicular traffic in the work area.

An outstanding reference for a project safety engineer's library is Suprenant and Basham's *Excavation Safety.*[4]

## REFERENCES

1. Richardson, Harold W. and Mayo, Robert S., *Practical Tunnel Driving,* McGraw-Hill Book Company, New York, 1960.
2. MacCollum, David V., "Systems safety and tunnel support," *National Safety News,* Dec. 1976.
3. American National Standards Institute, *Manual on Uniform Traffic Control Devices for Streets and Highways,* D6.1, U.S. Government Printing Office, Washington, DC.
4. Suprenant, Bruce A. and Basham, Kim D., *Excavation Safety,* The Aberdeen Group, Addison, IL, 1993.

# Chapter 24

*Phase Planning for Concrete Formwork,
Placement, and Erection of Precast
Components*

When you build a new house, be sure to put a railing around the edge of the
roof. Then you will not be responsible if someone falls off and is killed.

(Deuteronomy 22:8)

One of the most important safety requirement for formwork is to provide
safe access. (See Chapter 15 on fall prevention.) Many textbooks on formwork
address only structural requirements and completely ignore the need for
workers to have safe access for placement of concrete. On many projects
workers are required to use safety lines and belts to keep them from falling
when erecting formwork and placing concrete. Working from such a pre-
carious position dramatically reduces their efficiency, as it takes longer for
them to do the work than if working from a stable platform with handrails.
Staging can be designed as a permanent part of the formwork, and as the
formwork is moved from one part of the job to another, the staging stays
with it, eliminating the need to rebuild it at another location. In 1957 I
authored a paper on "How Proper Scaffolding Cuts Costs,"[1] which computed
the substantial savings from reduced labor costs that resulted when staging
was built to provide safe access, and workers did not have to rely on safety
belts and safety lines for fall protection.

## GENERAL REQUIREMENTS FOR CAST-IN-PLACE CONCRETE

1. The planning and design of formwork and falsework should be in accordance with provisions of the American Concrete Institute, Publication ACI 347-78, "Recommended Practice for Concrete Formwork," and ANSI A10.9, "Safety Requirements for Concrete Construction and Masonry Work." OSHA §1926.703(a)(8)(i) states: "The design of the shoring should be prepared by a qualified engineer and the erected shoring should be inspected by an engineer qualified in structural design." A "qualified engineer," in general terms, would be a registered professional engineer who has specialized in structural design.

2. All formwork, falsework, structural shoring, and bracing should be designed, erected, braced, and maintained so that it will safely support all vertical and lateral loads that might be applied until such loads can be supported by the structure.

3. The supporting ground or completed construction should be of adequate strength to carry the vertical and lateral loads to be imposed.

4. Splices should be designed and constructed to prevent buckling and bending.

5. Diagonal bracing should be provided in vertical and horizontal planes to provide stiffness and to prevent buckling of individual members.

6. The design of formwork and falsework should be submitted for review to the construction manager.

7. All vertical lift forms should be provided with scaffolding or work platforms completely encircling the area of placement.

8. For stability, single post shores should be horizontally braced in both the longitudinal and the transverse directions, and diagonal bracing also should be installed. Such bracing should be installed as the shores are being erected.

9. Stripped forms and shoring should be removed and stockpiled promptly after stripping. Protruding nails, wire ties, and other form accessories that are not necessary to subsequent work should be pulled or cut, or other means should be taken to eliminate the hazard.

10. Workers should be supplied with and required to wear eye and/or face protection during all operations involving nailing into concrete.

## SAFETY PLANNING FOR VERTICAL SHORING

1. When temporary storage of reinforcing rods, material, or equipment on top of formwork becomes necessary, these areas should be strengthened to meet the intended loads.

2. The sills for shoring should be sound, rigid, and capable of carrying the maximum intended load.
3. All shoring equipment should be inspected prior to erection to determine that it is as specified in the shoring layout. Any equipment found to be damaged should not be used for shoring.
4. Erected shoring equipment should be inspected immediately prior to, during, and immediately after the placement of concrete. Any shoring equipment that is found to be damaged, displaced, or weakened should be immediately reinforced or reshored.
5. Reshoring should be provided to safely support slabs and beams after stripping or where such members are subjected to superimposed loads due to construction.

## SAFETY PLANNING FOR TUBULAR WELDED FRAME SHORING

1. Metal tubular frames used for shoring should not be loaded beyond the safe working load recommended by the manufacturer.
2. All locking devices on frames and braces should be in good working order, coupling pins should align in the frame or panel legs, pivoted cross braces should have their center pivot in place, and all components should be in a condition similar to that of their original manufacture.
3. In checking the erected shoring frames with the shoring layout, the spacing between towers and cross brace spacing should not exceed that shown on the layout, and all locking devices should be closed.
4. Devices for attaching the external lateral stability bracing should be fastened to the legs of the shoring frames.
5. All baseplates, shore heads, extension devices, or adjustment screws should be in firm contact with the footing sill and the form.

## SAFETY PLANNING FOR TUBE AND COUPLER SHORING

1. Couplers (clamps) should not be used if they are deformed or broken, or have defective or missing threads on bolts or other defects.
2. The material used for the couplers (clamps) should be of a structural type such as drop-forged steel, malleable iron, or structural grade aluminum. Gray cast iron should not be used.
3. In checking the erected shoring towers with the shoring layout, the spacing

between posts should not exceed that shown on the layout, and all interlocking of tubular members and tightness of couplings should be checked.

4. All baseplates, shore heads, extension devices, or adjustment screws should be in firm contact with the footing sill and form material, and should be snug against the posts.

## SAFETY PLANNING FOR SINGLE POST SHORING

1. All baseplates or shore heads of single post shores should be in firm contact with the footing sill and form material.
2. Wherever single post shores are used in more than one tier, the layout should be designed and inspected by a structural engineer.
3. When formwork is at an angle, sloping, or when the surface shored is sloping, the shoring should be designed for such loading.
4. Adjustment of single post shores to raise formwork should not be made after concrete is in place.
5. Fabricated single post shores and adjusting devices should not be used if heavily rusted, bent, dented, or rewelded, or if they have broken weldments or other defects. If they contain timber, they should not be used if the timber is split or cut, has sections removed, is rotted, or is otherwise structurally damaged.
6. All timber and adjusting devices to be used for adjustable timber single post shores should be inspected before erection.
7. All nails used to secure bracing or adjustable timber single post shores should be driven home and the point of the nail bent over if possible.
8. For stability, single post shores should be horizontally braced in both the longitudinal and the transverse directions, and diagonal bracing also should be installed. Such bracing should be installed as the shores are being erected.

## SAFETY PLANNING FOR FALSEWORK

Sometimes falsework is used to install precast girders to avoid the use of cranes. Falsework provides a support so that precast girders can be rolled or slid on top of it and pushed into place with a hydraulic ram, using rollers or grease. This type of work should be done by permit and under the supervision of a licensed professional engineer who has specialized in such activities. The permit, specifications, drawings, and work methods should be reviewed and approved by the construction manager.


## SAFETY PLANNING FOR SLIP FORMS

1. The steel rods or pipe in which the jacks climb or by which the forms are lifted should be designed specifically for that purpose. Such rods should be braced where not encased in concrete.
2. Jacks and vertical supports should be positioned in such a manner that the vertical loads are distributed equally and do not exceed the capacity of the jacks.
3. The jacks or other lifting devices should be provided with mechanical dogs or other automatic holding devices to provide protection in case of failure of the power supply or the lifting mechanism.
4. Lifting should proceed steadily and uniformly and should not exceed the predetermined safe rate of lift.
5. Lateral and diagonal bracing of the forms should be provided to prevent excessive distortion of the structure during the jacking operation.
6. During jacking operations, the form structure should be maintained in line and plumb.
7. All vertical lift forms should be provided with scaffolding or work platforms completely encircling the area of placement.


## SAFETY PLANNING FOR PRECAST TILT-UP CONSTRUCTION

Precast tilt-up operations require not only the use of a crane with the necessary lifting capacity but also the use of shoring to brace the panels on both sides until permanent connections are completed, so they will not fall. (See OSHA §1926.704.) A registered professional engineer should design and supervise the design and the installation of both the temporary shoring and permanent connectors. The construction manager should review and approve this work before it is undertaken. OSHA §1926.704, "Requirements for Precast Concrete," states:

(a) Precast concrete wall units, structural framing, and tilt-up wall panels shall be adequately supported to prevent overturning and to prevent collapse until permanent connections are completed.

(b) Lifting inserts which are embedded or otherwise attached to tilt-up precast concrete members should be capable of supporting at least two times the maximum intended load applied or transmitted to them.

(c) Lifting inserts which are embedded or otherwise attached to precast concrete members, other than the tilt-up members, shall be capable of supporting at least four times the maximum intended load applied or transmitted to them.

(d) Lifting hardware shall be capable of supporting at least five times the maximum intended load applied or transmitted to the lifting hardware.

(e) No employee shall be permitted under precast concrete members being lifted or tilted into position except those employees required for the erection of those members.

## SAFETY PLANNING FOR LIFT-SLAB CONSTRUCTION

Since the L'Ambiance building collapse in Bridgeport, Connecticut, in 1987, which took some twenty-eight lives, lift-slab erection operations have become controversial. OSHA §1926.705, "Requirements for Lift-Slab Operations," states:

In addition to the general requirements of this Subpart, employers engaged in lift-slab operations shall comply with the existing specific requirements for lift-slab operations which are prescribed in ANSI A10.9-1970. These requirements, found in Section 11 of the American National Standard Safety Requirements for Concrete Construction and Masonry Work, ANSI, A10.9-1970, are reprinted in the Appendix to this section. In addition, there are applicable requirements in §1926.305 (a) and (b) of the Construction Safety and Health Standards which apply to jacks in general, and jacks used specifically in lift-slab construction.

The cited appendix reads as follows:

11.1 General. The safety requirements and recommendations in Section 11 apply specifically to lift-slab construction operations.

11.2 Design and Planning. Lift-slab operations should be designed and planned by a qualified professional engineer or architect. Such plans and designs shall include detailed instructions and sketches indicating the prescribed method of erection.

11.3.1  Jacking equipment should not be loaded beyond its safe working capacity, and then threaded rods and other members that transmit loads to the jacks should have a minimum safety factor of 2.5. Jacking equipment used in lift-slab operations should meet the criteria in 11.3.1 through 11.3.4. [Note: ANSI has interpreted this provision to mean that the safety factor of 2.5 must be met for all jacking components such as jacks, threaded rods, lifting nuts, and lifting angles, as well as shearheads, columns, and footings.]

11.3.2  Jacks should be designed and installed so that they will not continue to lift when overloaded.

11.3.3  Jacks should be installed with a safety device which will enable them to continue to support the load in any position should the jack malfunction and lose its lifting ability.

11.3.4  The maximum number of manually-controlled jacks on one slab should be limited to 14, and in no event should the number be too great to permit the operator to maintain the slab level within specified tolerances.

11.4  Uniform Lifting. Jacking operations should be synchronized in such a manner as to insure even and uniform lifting of the slab. During lifting, all points of the slab support should be kept within one-half inch of that needed to maintain the slab in a level position. If leveling is automatically controlled, a device should be installed which will stop the operation when the one-half-inch tolerance is exceeded or when there is a malfunction in the jacking system. If level is maintained by manual controls, such controls should be located in a central location and attended by a trained operator while lifting is in progress.

11.5  Falling Hazard. No one should be permitted under the slab during jacking operations. [Note: ANSI has interpreted this provision as follows: "No one is permitted in the building during jacking operations except those employees required for the jacking operation and to secure slabs."]

OSHA §1926.305, "Jacks—Lever and Ratchet, Screw, and Hydraulic," states:

(b) Lift slab construction. (1) Hydraulic jacks used in lift slab construction should have a safety device which will cause the jacks to support the load in any position in the event the jack malfunctions. (2) If lift slabs are automatically controlled, a device should be installed which will stop the operation when the 1/2-inch leveling tolerance is exceeded.

(c) Blocking. When it is necessary to provide a firm foundation, the base of the jack should be blocked or cribbed. Where there is a possibility of slippage of the metal cap of the jack, a wood block should be placed between the cap and the load.

## SAFETY PLANNING FOR ASSEMBLING BUILDINGS OF PRECAST COMPONENTS

Some precast buildings are built one story at a time, a practice that limits the hazard of falls to approximately ten feet, which is better than a fall of many feet as can happen in other methods. Staging should be provided for workers while they are securing the precast beams to the pillars and placing the first floor panels for the next level. Well-anchored hatchcovers should be immediately placed over all floor openings for future elevators and duct work for pipes and conduits, to prevent falls from one level to another, and temporary stairways should be erected between floors. As the next floor is being set in place, perimeter guarding should be installed that has a toprail, midrail, and toeboard, with the toprail being able to resist 200 pounds of lateral force.

## SAFETY PLANNING FOR USE OF CRANES IN CONCRETE WORK

All cranes should be examined for safety, as outlined in Chapter 14, before they are used. Determining if the crane selected has the capacity to lift the required loads at the intended radii is an important part of safety planning. Signalers must be well trained and must be strategically positioned so that no matter which direction the boom is slued, they always have a clear view of the crane operator and the pickup and placement areas. A radio or telephone communication system should be used when the operator cannot see the location where the load must be placed. A crane never should be used in a location where its boom can reach into powerlines. The swing area of the crane cab should be barricaded to prevent the entry of workers into this danger zone.

## SAFETY PLANNING FOR USE OF OTHER EQUIPMENT

1. Bulk storage bins, containers, or silos should have conical or tapered bottoms with mechanical or pneumatic means of starting the flow of material.

2. Concrete mixers equipped with one-yard or larger loading skips should be equipped with a mechanical device to clear the skip of material.

3. Mixers of one-yard or greater capacity should be equipped with protective guardrails installed on each side of the skip.

4. Handles on bull floats used where they may contact energized electrical conductors should be constructed on nonconductive material, or insulated with a nonconductive sheath whose electrical and mechanical characteristics provide the equivalent protection of a handle constructed of nonconductive material. (See Chapter 22 for removal of the powerlines from the construction site.)

5. Powered and rotating concrete troweling machines that are manually guided should be equipped with a control switch that will automatically shut off the power whenever the operator removes his/her hands from the equipment handles.

6. Handles of buggies should not extend beyond the wheels on either side of the buggy. Installation of knuckle guards on buggy handles is recommended.

7. Pumpcrete or similar systems using discharge pipes should be provided with pipe supports designed with a safety factor of two. The compressed air hose in such systems should have positive fail-safe joint connectors to prevent separation of sections when pressurized.

8. Concrete buckets equipped with hydraulic or pneumatically operated gates should have positive safety latches or similar safety devices installed to prevent premature or accidental dumping. The buckets should be designed to prevent excess aggregate and loose material from accumulating excessively on the top and the sides of the bucket.

9. Planning should ensure that vibrator crews will be kept out from under concrete buckets suspended from cranes or cableways.

10. For discharging on a slope, the wheels of ready-mix trucks should be blocked and the brakes set to prevent movement.

11. Nozzle operators applying a cement, sand, and water mixture through a pneumatic hose should wear protective hand and face equipment.

REFERENCE

1. MacCollum, David V., "How proper scaffolding cuts costs," *Western Construction,* Sept. 1957.

# Chapter 25

## Phase Planning for Steel Erection

If a builder build a house for a man and do not make its construction firm and the house which he has built collapse and cause the death of the owner of the house—that builder shall be put to death.
(Code of Laws of Hammurabi, King of Babylonia, 1792–1750 B.C.

The way in which a steel-framework building is designed can substantially reduce the hazard of "falls-from-an-elevation." The preassembling of structural components such as trusses, framing, and reinforcing steel, on the ground and using a large crane to lift them into place, eliminates the need to connect structural members at elevated locations and is an excellent method of fall prevention. Ironworkers sustain the highest percentage of falls-from-an-elevation when compared to other crafts, and some 60 percent of their fatalities are the result of falling from a steel structure. There is a continuing need for fall protection during steel erecting because, in many instances, a building cannot be erected without the ironworkers' assembling it aloft. (See Figures 25-1 and 25-2.) The best book on fall protection is Ellis's *Introduction to Fall Protection.*[1] Chapter 15 of this book also addresses fall prevention in some detail.

ANSI Standard A10.13, "Steel Erection Safety Requirements," is the most thorough reference for steel erection. The Corps of Engineers' *Safety and*

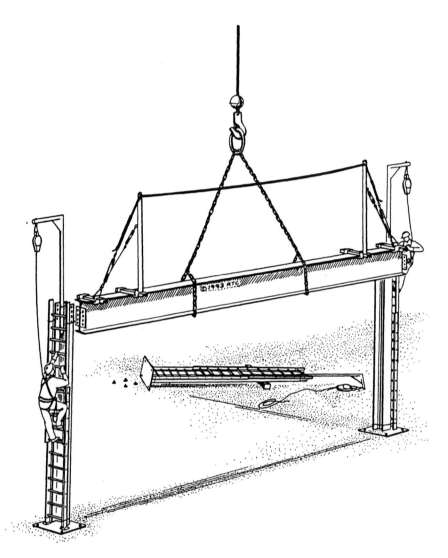

**Figure 25-1.**    Fall protection, steel beam. (Courtesy of the Research and Trading Corporation.)

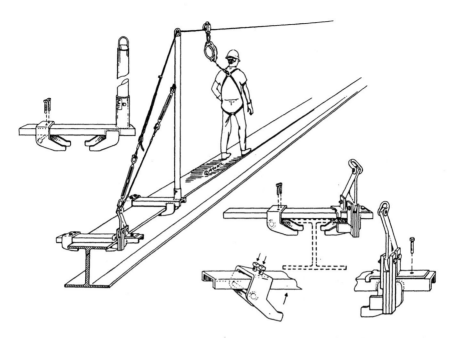

**Figure 25-2.** Fall protection, steel beam alternate. (Courtesy of the Research and Trading Corporation.)

*Health Requirements Manual,* EM 385-1-1,[2] also is an excellent reference. Many of the following guidelines have been taken from these two references.

## SAFETY PLANNING FOR PERMANENT FLOORING

1. The permanent floors should be installed as the erection of structural members progresses, and there should be not more than eight stories between the erection floor and the uppermost permanent floor, except where the structural integrity is maintained by design.
2. At no time should there be more than two floors or thirty feet of unfinished bolting or welding above the foundation or the uppermost permanently secured floor unless the column is one continuous member and approval

has been obtained from the designated authority. In no case should four floors or forty-eight feet be exceeded.

## SAFETY PLANNING FOR TEMPORARY FLOORING

1. The derrick or erection floor should be planked or decked over its entire surface except for access openings. Planking or decking should be of a sufficient thickness to carry the working load. Planking should be not less than two inches thick full size undressed, and should be laid tight and secured to prevent movement.
2. On structures not adaptable to temporary floors, and where scaffolds are not used, safety nets should be installed and maintained whenever the potential fall distance exceeds two stories or twenty-five feet.
3. Where skeleton steel erection is being done, a tightly planked and substantial floor should be maintained within two stories or thirty feet, whichever is less, directly under that portion of each tier of beams on which any work is being performed, except for gathering and stacking of temporary floor planks on a lower floor in preparation for transferring such planks to an upper floor. Where such a floor is not practical, safety nets should be required.
4. In gathering and stacking of temporary floor planks, the planks should be removed successively, with employees working toward the last panel of the temporary floor so that the work is always done from a planked floor.
5. In gathering and stacking of temporary floor planks from the last panel, the employees assigned to such work should be protected by an approved fall protection system.

## SAFETY PLANNING FOR OTHER FLOORING

1. In the erection of a building having double wood floor construction, the rough flooring should be completed as the building progresses, including the tier below the one on which floor joists are being installed.
2. For single wood floor or other flooring systems, the floor immediately below the story where the floor joists are being installed should be kept planked or decked over.

## SAFETY PLANNING FOR STRUCTURAL STEEL ASSEMBLY

1. During the final placing of solid web structural members, the load should not be released from the hoisting line until the members are secured with not less than two bolts or 10 percent of the bolts, whichever is greater, at each connection, and drawn up wrench-tight.
2. Open web steel joists should not be placed on any structural steel framework until it is permanently bolted or welded.
3. In steel framing, where bar joists are utilized and columns are not framed in at least two directions with structural steel members, a bar joist should be field-bolted at columns to provide lateral stability during construction.
4. When longspan joists or trusses, forty feet or longer, are used, a center row of bolted bridging should be installed to provide lateral stability during construction prior to slacking of the hoisting line.
5. No load should be placed on open web steel joists until these stability requirements are met.
6. Structural and reinforcing steel for walls, piers, columns, and similar vertical structures should be guyed and supported to prevent collapse.
7. When bolts, drift pins, or rivets are being knocked out, they should be kept from falling.
8. Impact wrenches should have a locking device for retaining the socket.
9. Connections of equipment used in plumbing-up should be secured.
10. The turnbuckles should be secured to prevent unwinding while under stress.
11. Plumbing-up guys and related equipment should be placed so that workers can get at the connection points.
12. Plumbing-up guys should be removed only under the supervision of a competent person.
13. Wire mesh, plywood, or the equivalent should be used around columns where planks do not fit tightly.
14. Temporary flooring should be secured against displacement.
15. Riveting should not be done near combustible material unless precautions are taken to prevent fire.
16. Containers should be provided for storing or carrying rivets, bolts, and drift pins, and secured against accidental displacement when aloft.
17. A safety wire should be properly installed on the snap and on the handle of the pneumatic riveting hammer and should be used at all times. The wire size should be not less than No. 9 (B&S gauge) leaving the handle and annealed No. 14 on the snap or its equivalent.

## SAFETY PLANNING FOR USE OF CRANES

All cranes should be examined for safety (as outlined in Chapter 14) before they are used. Determining if the crane selected has the capacity to lift the required loads at the intended radii is an important part of safety planning. Another very important consideration is the length of the boom. It must be long enough to be able to lift the last few steel beams safety into place on the top of the structure. If the boom is not long enough, two-blocking of the hoist line can occur, and the load can be dropped. Signalers must be well trained and must be strategically positioned so that no matter which direction the boom is slued, they always have a clear view of the crane operator and the pickup and placement areas. A radio or telephone communication system should be used when the operator cannot see the location where the load must be placed. A crane should never be used in a location where its boom can reach into powerlines. The swing area of the crane cab should be barricaded to prevent entry of workers into this danger zone.

## REFERENCES

1. Ellis, J. Nigel, *Introduction to Fall Protection,* 2nd Edition, American Society of Safety Engineers, 1994.
2. *Safety and Health Requirements Manual,* EM 385-1-1, U.S. Army Corps of Engineers.

# Chapter 26

## Phase Planning for Mechanical Installation

The tank rested on the backs of twelve bronze bulls that faced outwards, three facing in each direction.
(2 Chronicles 4:4, building of King Solomon's temple)

Subcontractors installing plumbing, elevators, and systems for air conditioning and fire prevention usually carry out their various activities in a common work area, and the activities of one subcontractor can endanger the work of other subcontractors or crafts. The initial phase planning should schedule work to avoid conflicts that create hazardous situations. The superintendent in charge of a particular phase and the project safety engineer should evaluate the work in progress just before a new contractor is to start work to see if any conflicts will arise that might create danger for the new contractor or those already on the job. If they see any dangerous conflicts, they should advise the construction manager of the danger and what needs to be done. The construction manager should convene a short meeting with the new contractor and the ones already on the job so all problem areas will be resolved.

Because of the variety of work that goes on from day to day, hazardous situations can develop very quickly. The main thrust of phase planning should be to see that a system of rapid communication between managers of the various subcontractors is established. Sometimes a daily five-minute meeting to review what each one plans will suffice.

Several things need to be considered when subcontractors share a common work area:

1. Does the work create a hazard for others?
2. Can the work be scheduled when others are not in the same work area?
3. Is there a common use of staging, aerial lifts, ladders, and other access systems, and are they adequate?
4. Is a lockout system needed to avoid injury to others?

Often the main objective of a subcontractor is to get the equipment installed, see that it works, pass the acceptance test, and be gone. In order to test the equipment, the subcontractor may have to energize part of the facility's permanent electrical system. Other subcontractors may be installing other equipment connected to the same power source; so they must be forewarned that the system is going to be energized. The construction manager and the project safety engineer must make sure that a safety plan is in place before such tests are made. This includes a written lockout procedure available to all parties and a method of informing everyone of the power sources affected, the physical movement of the equipment under test (elevators, conveyors, etc.), the length of time the test will take, and when will it be completed. The manufacturer of the machinery or the equipment to be tested should have detailed instructions on how the test should be conducted and should be able to provide excellent guidelines for developing a safety plan for acceptance testing. When pieces of mechanical equipment are left over and not installed, watch out! Something is probably amiss. Ask why the parts are left over. There may be a good reason, but maybe not. There might have been an unintentional oversight. Manufacturer's reps usually are worth their weight in gold during test and start-up. The construction manager should contact them several weeks prior to installation, start-up, and acceptance testing so that they can assist in safety planning.

Chapter *27*

## *Phase Planning for Electrical Systems*

Evidently we have to have a great many tremendous kicks in the pants before we learn anything.

(Aldous Huxley)

Electrical systems, both temporary and permanent, affect everybody on a construction project. As mentioned several times, the danger created by powerlines to those operating cranes and other high-clearance equipment is one of the first things to consider in safety planning. Because every contractor on a job will need to use electrical power, other hazards associated with the use of electrical systems also must be examined.

The electrical system should be planned so that a single party or a subcontractor is responsible to the construction manager for the entire electrical system through the life of the project, to maintain continuity, as those who install it are the ones most familiar with it and its relatonship to those who use it. Centralized control makes it easier to modify the system to provide more power for special activities. The supervisor in charge of the electrical system and the safety engineer should jointly review upcoming work that may create dangerous situations and advise the construction manager of any anticipated hazards so that plans can be modified and appropriate parties notified.

The issuance of safety clearances and permits for work requiring lockouts, deenergizing electrical systems, or working with live lines must be under the supervision and the scrutiny of the construction manager to ensure that appropriate coordination will be made with the power company, as outlined in Chapters 22 and 31.

In addition to electrical power systems, other systems for communication, control, and emergency alarm will be installed by various contractors. The basic procedures used to indoctrinate these subcontractors are the same as those outlined in Chapter 26.

Temporary and permanent electrical systems often interface and are intertwined during construction activities, and should be reviewed during initial safety planning and phase planning.

The *National Electrical Safety Code* (NESC), the *National Electrical Code* (NEC), OSHA, and the Corps of Engineers' EM 385-1-1 give excellent guidance for electrical phase planning, and have been excerpted in the following pages.

## SAFETY PLANNING FOR ELECTRICAL WIRING AND APPARATUS

1. All electrical wiring and equipment should be a type listed by Underwriters' Laboratories (UL), the Factory Mutual Engineering Corporation, or another recognized test or listing agent for the specific application.
2. All high-voltage distribution installations should comply with the NESC, low-voltage systems with the NEC, and marine installations with U.S. Coast Guard regulations.
3. All work should be done by personnel familiar with code requirements and qualified for the class of work to be performed.
4. Live parts of wiring or equipment should be guarded to protect all persons or objects from harm.
5. Transformer banks or high-voltage equipment should be protected from unauthorized access. Entrances not under constant observation should be kept locked. Signs warning of high voltage and prohibiting unauthorized entrance should be posted at entrances. Metallic enclosures should be grounded.
6. Gates or doors to enclosures for electrical equipment should swing outward or provide clearance from installed equipment.
7. Electrical wire or flexible cord passing through work areas should be covered or elevated to protect it from damage by foot traffic, vehicles, sharp corners, projections, or pinching.
8. Before work is begun, the person in charge should ascertain by inquiry, direct observation, or instruments whether any part of an electric power

circuit, exposed or concealed, is so located that the performance of the work may bring any person, tool, or machine into physical or electrical contact therewith. Whenever possible, all equipment as well as circuits to be worked on should be deenergized before work is started, and personnel should be protected by clearance procedures and grounding.

9. When it is necessary to work on energized lines and equipment, rubber gloves and other protective equipment or hotline tools meeting the provisions of the ANSI J-6 series should be used.

10. At least two persons should be assigned to work on energized overhead lines, in substations and power plants where the wiring is congested, where the work is at remote or isolated locations, at night, during inclement weather, or when handling energized conductors or apparatus. One person, trained to recognize the electrical hazards, should be delegated to watch the movements of the other person(s) doing the work so that he/she can warn them if they get dangerously close to live conductors or perform other unsafe acts, and assist in case of an accident.

11. Insulation mats or platforms of substantial construction and providing good footing should be so placed on floors and on the frames of equipment having exposed live parts that the operator or persons in the vicinity cannot touch such parts unless standing on the mats, platforms, or insulated floors.

12. All circuits should be protected against overload.

13. Suitable barriers or other means should be provided to ensure that the work space for electrical equipment should not be used as a passageway when energized parts of electrical equipment are exposed.

14. No overcurrent device or other disconnect should be placed in any permanently grounded conductor, except where the device simultaneously opens all conductors of the circuit.

15. When fuses are installed or removed with one or both terminals energized, special tools insulated for the voltage should be used.

16. Standard plugs and receptacles for a particular voltage, amperage, frequency, or type of current as specified in National Electrical Manufacturers Association (NEMA) standards should be used.

## SAFETY PLANNING FOR DISCONNECT AND OVERCURRENT PROTECTION

1. Switches, fuses, and automatic circuit breakers should be marked, labeled, or arranged for ready identification of circuits or equipment supplied through them.

2. Switches should be of the enclosed safety type, with enclosures grounded, and installed so as to minimize the danger of accidental operation.
3. Fuse cabinets should have close-fitting doors that can be locked.
4. Switches or breakers should be provided with a means for locking them in the off-position during maintenance periods.
5. Fuses and circuit breakers should be of the proper rating for the circuit protected.
6. A readily accessible, manually operated switch should be provided for each incoming service or supply circuit.
7. Switches, circuit breakers, fuse panels, and motor controllers in wet locations or outside should be in a weatherproof enclosure or cabinet.
8. Disconnecting means should be so located or shielded that persons will not be injured when the disconnect is operated.
9. Overcurrent protection devices must be readily accessible, clearly labeled, not exposed to physical damage, and not placed in the vicinity of easily ignitable materials.

## SAFETY PLANNING FOR GROUNDING

1. All electrical circuits should be grounded in accordance with the NEC and the NESC unless otherwise noted in this manual.
2. A ground should be provided for non-current-carrying metallic parts of such equipment as generators (if not exempted by NEC 250-6), electrically powered arc welders, switches, motor controller cases, fuse boxes, distribution cabinets, frames, non-current-carrying rails used for travel and motors of electrically operated cranes, electric elevators, metal frames of nonelectric elevators to which electric conductors are attached, other electric equipment, and metal enclosures around electric equipment.

## SAFETY PLANNING FOR TEMPORARY WIRING

1. Temporary powerlines, switch boxes, receptacle boxes, metal cabinets, and enclosures around equipment should be marked to indicate the maximum operating voltage.

2. Temporary wiring should be guarded, buried, or isolated by elevation to prevent accidental contact by workers or equipment.

3. Vertical clearance above walkways should not be less than ten feet from circuits carrying 600 volts or less.

4. Wires with nonrated weatherproof insulation should not be enclosed in metal raceways or used for wiring in tanks, penstocks, and tunnels. Receptacles used in damp or wet locations should be approved for the purpose. Where a receptacle is installed outdoors (outdoors is considered a wet location), it should be contained in a weatherproof enclosure, the integrity of which is not affected when an attachment plug is inserted.

5. Wires should be insulated from their supports.

6. All temporary lighting strings in outdoor or wet locations such as tunnels, culverts, valve pits, outlets, floating plants, and so on, should consist of lamp sockets and connection plugs permanently molded to the hard service cord insulation.

7. Temporary wiring installed in conduit should have bushings at all outlets and terminals.

8. Flexible cord sets should be listed by UL or another recognized listing agent. Flexible cord sets used on construction sites should contain the number of conductors required for the service plus an equipment ground wire. The cords should be hard usage or extrahard usage as specified in the NEC. Approved cords may be identified by the word "outdoor" or the letters "WA" on the jacket.

9. Portable and semiportable electrical tools and equipment should be grounded by a multiconductor cord having an identified grounding conductor and a multicontact polarized plug-in receptacle.

10. Ground fault circuit interrupters (GFCIs) are required in all circuits used for portable electric tools to include block/brick saws. The GFCI should be calibrated to trip within the threshold values of 5 ma + 1 ma as specified in UL Standard 943. All GFCIs should be UL-listed and installed in accordance with the most recent edition of the *National Electric Code*. The permanent wiring should be electrical circuits grounded in accordance with the NEC. GFCIs may be sensitive to some equipment such as concrete vibrators. In these instances, an assured equipment grounding conductor program is acceptable. This program should be IAW 29 CFR §1926.404.

11. Semiportable equipment, floodlights, and work lights should be grounded. The protective ground of such equipment should be maintained during moving unless supply circuits are deenergized.

12. Tools protected by an approved system of double insulation, or its equivalent, need not be grounded. Double-insulated tools should be distinctly marked and listed by UL or FM.

13. Grounding rods must be at least ⅝-inch-diameter steel or iron rods, ½-inch-diameter copper-clad steel, or ¾-inch-diameter galvanized pipe. They should be in unbroken eight-foot lengths and driven to full depth. Resistance to ground should not exceed twenty-five ohms.

14. Grounding circuits should be checked to ensure that the circuit between the ground and a grounded power conductor has a resistance low enough to permit a current flow sufficient to cause the fuse or the circuit breaker to interrupt the current.

15. Conductors used for bonding and grounding stationary and movable equipment should be of ample size to carry the anticipated current. When one is attaching bonding and grounding clamps or clips, a secure and positive metal-to-metal contact should be made. The ground end should be attached and removed by insulated tools or other suitable devices. When one is removing grounds, the grounding device first should be removed from the line or the equipment by using insulated tools or other suitable devices. Such bonding and grounding attachments should be made before closures are opened and material movements are started, and they should not be broken until after material movements are stopped and closures are made.

16. Attachment plugs or other connectors supplying equipment at more than 300 volts should be skirted or otherwise designed so that arcs will be confined.

17. Attachment plugs for use in work areas should be constructed so that they will endure rough use and be equipped with a cord grip to prevent strain on the terminal screws.

18. Flexible cord should be used only in continuous lengths without splices, except that molded or vulcanized splices may be used if made by a qualified electrician. The insulation should be equal to the cable being spliced, and wire connections will be soldered.

19. Patched, oil-soaked, worn, or frayed electric cords or cables should not be used.

20. Portable handlamps should be of the molded composition or other type approved for the purpose. Metal-shell, paper-lined lampholders should not be used. Handlamps should be equipped with a handle and a substantial guard over the bulb that is attached to the lampholder or the handle.

21. Extension cords or cables should not be fastened with staples, hung from nails, or suspended by bare wire.

22. Disconnecting boxes should be securely fastened to the surface and fitted with covers.

23. Splices in trailing cable should be mechanically strong and insulated to retain the mechanical and the dielectric strength of the original cable.

24. Plugs and receptacles should be kept out of water unless of an approved submersible type.
25. Bulbs attached to temporary lighting strings and extension cords should be protected by wire guards or the equivalent unless deeply recessed in a reflector. Temporary lights should not be suspended by their electric wire unless the cord and the lights are designed for this suspension.
26. When temporary wiring is used in tanks or other confined spaces, an approved switch, identified and marked, should be provided at or near the entrance to such spaces for cutting off the current in emergencies.
27. Exposed empty light sockets and broken bulbs should not be permitted.
28. Portable electric lighting used in confined wet and/or hazardous locations such as drums, tanks, vessels, and grease pits should be operated at a maximum of twelve volts.
29. Temporary electrical distribution systems and devices should be checked and accepted for polarity, ground continuity, and ground resistance prior to initial use and prior to use after modification. The ground resistance should be measured, recorded, and furnished to the designated authority at the time of installation and should comply with item 14 above.
30. Contractors should submit a sketch of proposed temporary power distribution systems before temporary power is installed.
31. Nonmetallic sheathed cable may be used as allowed by the NEC and as follows:
    a. Along studs, joists, or similar supports closely following the building finish or running boards when 7′8″ or more above the floor.
    b. When firmly attached to each cabinet, box fitting, or fixture by means of a cable clamp.
32. Nonmetallic sheathed cable may not be used where precluded by the NEC, nor may it be used as follows:
    a. As portable extension cords.
    b. Lying on the ground subject to any type of traffic.
    c. Where subject to frequent flexing.
    d. As service entrance cable.

## SAFETY PLANNING FOR OPERATIONS ADJACENT TO OVERHEAD LINES

1. Overhead transmission and distribution lines should be carried on towers and poles that provide safe clearances over roadways and structures.
2. All electrical power or distribution lines should be placed underground in

areas where there is extensive use of equipment having the capability of encroachment on safe clearance distances, as specified in item 7 below.

3. Clearances should be adequate for the movement of vehicles and for the operation of construction equipment.

4. Protection of outdoor trolleys and portable cables rated above 600 volts for supplying power to movable construction equipment such as gantry cranes, mobile cranes, shovels, and so on, should conform to the NESC.

5. When it is necessary to transport machinery or equipment under overhead lines in a manner that encroaches on specified clearances, the job should be scheduled so that the lines can be deenergized.

6. Operations adjacent to overhead lines should not be initiated until coordinated with the utility officials.

7. Operations adjacent to overhead lines are prohibited unless one of the following conditions is satisfied:

   a. Power has been shut off, and positive means have been taken to prevent the lines from being energized.

   b. The minimum clearance from energized overhead lines is as shown in Table 27-1, or the equipment will be positioned and blocked to ensure that no part, including cables, can come within the minimum clearances shown in Table 27-1.

**Table 27-1.** Minimum clearances from powerlines

| Powerlines nominal system kv | Minimum required clearance, feet |
|---|---|
| 0–50 | 10 |
| 51–100 | 12 |
| 101–200 | 15 |
| 201–300 | 20 |
| 301–500 | 25 |
| 501–750 | 35 |
| 751–1000 | 45 |

8. Chapter 22 includes much information on keeping powerlines and cranes and other high-clearance equipment separated.

## SAFETY PLANNING FOR LIGHTING

1. Emergency entrances and exits should be illuminated with lighting to provide a minimum of one foot-candle, measured at the floor, and should

be arranged so that the failure of any single lighting unit, including the burning out of an electric bulb, will not leave any area in total darkness.

2. While work is in progress, offices, facilities, accessways, working areas, construction roads, and so on, should be lighted by at least the minimum light intensities specified in Table 27- 2.

**Table 27-2.**   Minimum light intensities

| Location | Foot-candles |
|---|---|
| Accessways: | |
| General indoor | 5 |
| General outdoor | 3 |
| Exitways, walkways, ladders, stairs | 10 |
| Administrative areas (offices, drafting, meeting rooms, etc.) | 50 |
| Construction areas | |
| General indoor | 5 |
| General outdoor | 3 |
| Tunnels and general underground work areas (minimum of 10 foot-candles required at tunnel and shaft heading during drilling, mucking, and scaling) | 5 |
| Conveyor routes | 10 |
| Docks and loading platforms | 3 |
| Elevators, freight and passenger | 20 |
| First aid stations and infirmaries | 30 |
| Maintenance, operating, and construction areas | |
| Vehicle maintenance shop | 30 |
| Carpentry shop | 10 |
| Outdoors field maintenance area | 5 |
| Refueling area, outdoors | 5 |
| Shops, fine detail work | 50 |
| Shops, medium detail work | 30 |
| Welding shop | 30 |
| Mechanical/electrical equipment rooms | 10 |
| Parking areas | 3 |
| Toilets, wash and dressing rooms | 10 |
| Warehouses and storage rooms and areas | |
| Stockrooms, active or bulk storage, indoors | 10 |
| Inactive storage, indoors | 5 |
| Rack storage, indoors | 25 |
| Outdoor storage | 3 |
| Work areas not listed | 30 |

# Chapter 28

## *Phase Planning for Interiors*

The inside walls were covered with cedar panels from floor to ceiling and the floor was made of pine.

(1 Kings 6:15, building of King Solomon's temple)

Safe access to high-rise structures is most critical when workers are completing building interiors. A temporary outside elevator or a material-hoist often is used for this purpose. Because the presence of a material-hoist gives people the incentive to use it as a personnel lift, the best choice would be to provide a temporary elevator suitable for lifting both people and materials.

If elevators are to be used, the following requirements, taken from the Corps of Engineers' EM 385-1-1, should be considered.

## SAFETY PLANNING FOR SAFE PASSENGER AND FREIGHT ELEVATORS

1. Permanent elevators should be designed, constructed, inspected, tested, and maintained in accordance with applicable provisions of ANSI A17.1, "Safety Code for Elevators, Dumbwaiters, and Moving Walks," and ANSI A17.2, "Practice for Inspection of Elevators." Personnel hoists installed

inside or outside buildings during construction, alteration, or demolition should be designed, constructed, inspected, and operated in accordance with ANSI A10.4, "Safety Requirements for Personnel Hoists." Temporary hoists used to raise and lower persons during construction demolition operations should be designed, constructed, inspected, and operated in accordance with ANSI A10.22, "Safety Requirements for Rope-Guided and Nonguided Workmen's Hoists."

2. Positive protection should be provided to prevent persons or objects from falling into the shaftway. Personnel hoist towers outside the structure should be enclosed for the full height on the side or sides used for entrance and exit for the structure. At the lowest landing, the enclosure on the sides not used for entrance and exit should be enclosed to at least ten feet. Other sides of the tower adjacent to floors or scaffold platforms should be enclosed ten feet above the level of such floors or scaffolds. Personnel hoist towers inside of structures should be enclosed on all four sides throughout the full height.

3. Safe accessways should be provided for servicing and inspection of hoist towers and equipment.

4. Cars should be permanently enclosed on all sides and the top except for sides used for entrance and exit, which have car gates or doors.

5. A door or a gate should be provided at each entrance to the car, which should protect the full width and height of the car entrance opening.

6. Doors or gates should be provided with electric contacts that do not allow movement of the hoist when a door or a gate is open.

7. Overhead protective covering of two-inch planking, ¾-inch plywood, or another solid material of equivalent strength should be provided on the top of every personnel hoist.

8. Safeties should be capable of stopping and holding the car and the rated load when traveling at governor tripping speed.

9. The car should be provided with a capacity and data plate secured in a conspicuous place on the car or the crosshead.

10. Internal combustion engines should not be permitted for direct drive.

11. Normal and final terminal stopping devices should be provided.

12. An emergency stop switch should be provided in the car and marked "Stop."

13. The minimum number of hoisting ropes used should be three for traction hoist and two for drum-type hoists.

14. The minimum diameter of hoisting and counterweight wire ropes should be ½-inch.

15. Minimum safety factors for suspension wire ropes are as shown in Table 28-1.

**Table 28-1.**   Safety factors for suspension wire ropes

| Rope speed in feet per minute | Minimum factor of safety |
|---|---|
| 50 | 7.60 |
| 75 | 7.75 |
| 100 | 7.95 |
| 125 | 8.10 |
| 150 | 8.25 |
| 175 | 8.40 |
| 200 | 8.60 |
| 225 | 8.75 |
| 250 | 8.90 |
| 300 | 9.20 |
| 350 | 9.50 |
| 400 | 9.75 |
| 450 | 10.00 |
| 500 | 10.25 |
| 550 | 10.45 |
| 600 | 10.70 |

16. Personnel hoists used in bridge tower construction should be approved by a registered engineer and erected under the supervision of a qualified engineer competent in this field.
17. Personnel hoists should be inspected and maintained on a weekly basis. Whenever the hoisting equipment is exposed to winds exceeding 35 mph, it should be inspected and put in operable condition before reuse.
18. Wire rope should be taken out of service when there are broken wires or the rope shows excessive wear, corrosion, or defect.
19. Rated load capacities, recommended operating speeds, and special hazard warnings and instructions should be posted on cars and platforms.
20. Endless belt manlifts should be prohibited on construction projects.

## SAFETY PLANNING FOR SAFE USE OF MATERIAL HOISTS

1. Material hoists must never be used by people.
2. Material hoists should be designed, constructed, inspected, tested, and

maintained in accordance with the "Safety Requirements for Material Hoists," ANSI A10.5.

3. Towers, shaftways, and cars of material hoists should be constructed of sound material capable of sustaining the maximum load to be imposed with a factor of safety of at least five.

4. Cars or platforms of all hoisting equipment should be equipped with broken cable safety devices.

5. All towers should rest on solid foundations, be plumb, and be well guyed or otherwise strongly anchored in four directions at the top and at least every thirty feet in height.

6. Hoist platforms or cars should have all unused sides enclosed, and the roof should be constructed of material strong enough to afford protection from falling objects.

7. The roof may be hinged in sections to provide for the hoisting of long material.

8. Hoist towers should be enclosed on all sides for their entire height with a wire screen enclosure, formed of not less than No. 18 U.S. gauge wire with openings not exceeding ½-inch mesh securely fastened to the tower structure, with openings formed onto each floor level.

9. Not more than one cage or bucket should be operated at the same time by any one hoisting machine or operator.

10. Each electric-motor-driven hoist should have an electromechanical automatic motor brake or an electrical device to automatically hold the load in case of power failure.

11. Operating rules should be established and posted at the operator's station of the hoist. Such rules should include the signal system and the allowable line speed for various loads. Rules and notices should be posted on the car frame or crosshead in a conspicuous location, including the words "No Riders Allowed."

12. No person should be allowed to ride on material hoists except for inspection and maintenance.

13. All entrances of the hoistways should be protected by substantial gates or bars, which should guard the full width of the loading entrance. All hoistway entrance bars and gates should be painted with diagonal contrasting colors, such as black and yellow stripes.

14. Bars at hoistway entrances should not be less than 2″ × 4″ wooden bars, or the equivalent, located two feet from the hoistway line. Bars should be located not less than thirty-six inches or more than forty-two inches above the floor.

15. Gates or bars protecting the entrances to hoistways should be equipped with a latching device.

16. The operator's station of a hoisting machine should be provided with overhead protection equivalent to tight planking not less than two inches thick. The support for the overhead protection should be of equal strength to that of the protection.
17. All material hoist towers should be designed by a registered engineer competent in this field.

## SAFETY PLANNING FOR THE USE OF OTHER LIFTING DEVICES

Among the many questions that should be considered if materials are to be lifted to upper levels by forklifts, loaders, or small truck-mounted cranes are the following:

1. Are there any adjacent powerlines?
2. Is the ground level where the lifting equipment will be positioned?
3. Will the ground support the weight of the equipment?
4. To whom does the equipment belong?

Sometimes a supplier will deliver building materials to the job site on a truck with a pedestal on its flatbed, which is not always suitable. If the crane is not fitted with a load-moment indicator (LMI), operators sometimes can have a false sense of stability when lifting over either the rear or the front, as the rated capacity may only be 40 to 50 percent of the tipping load in these positions, and this can cause an upset or cause the boom to buckle.

Remember, as mentioned in Chapter 15, that suppliers sometimes fail to replace handrails on upper levels properly after the delivery of materials; so be sure to check for this hazard.

## SAFETY PLANNING TO ELIMINATE DANGEROUS FLOOR AND WALL OPENINGS

Inside the building, the danger of falls through floor openings and from outside wall openings is the greatest during the completion of the interior. The Corps of Engineers' EM 385-1-1 gives good guidance.

1. A floor or roof "hole" is an opening measuring less than twelve inches but more than one inch in its least dimension; a floor or roof "opening" is an opening measuring twelve inches or more in its least dimension. A wall "hole" is an opening less than thirty inches but more than one inch in height and of unrestricted width; a wall "opening" is an opening at least thirty inches high or higher and at least eighteen inches wide. Holes should be covered primarily to prevent workers from stepping into or through them, and materials from falling into or through them. Openings should be covered or guarded to prevent people or materials from falling through them.

2. All floor and roof holes, skylights, and openings into which people can accidentally walk should be guarded by an enclosure guard or covered with material and bracing of sufficient strength to support any load that may be imposed on them. Coverings for floor and roof openings should be secured in place to prevent their accidental removal or displacement. Standards for guardrailing, intermediate rails, and toeboards should be as specified.

3. Every stairway and ladderway floor opening should be guarded on all exposed sides except the entrance opening by a securely anchored standard railing with intermediate rail and toeboard.

4. Temporary stairway and ladderway entrances should be offset or provided with a gate to prevent anyone from walking into the opening.

5. Every hatchway and chute floor opening should be guarded by a hinged floor-opening cover equipped with railings attached so as to leave only one exposed side. The exposed side should be either provided with a swinging gate or so offset that a person cannot walk into the opening.

6. Every opening from which there is a drop of more than four feet and whose bottom is less than three feet above the working surface should be guarded with a top rail, top rail and intermediate rail, or standard guardrail. A toeboard should be provided where the bottom of the wall opening, regardless of width, is less than four inches above the working surface.

7. An open-sided floor or platform six feet or more above the adjacent floor or ground level should be guarded by a standard guardrail or the equivalent.

8. Doors or gates opening into a stairway should have a platform, and swinging of the door should not reduce the width of the platform to less than twenty inches.

9. A chute opening into which debris is dumped should have a guardrail approximately forty-two inches above the floor or other surface on which workers stand to dump the material.

10. An extension platform outside a wall opening into which materials can be hoisted for handling should have a standard railing. However, one side of

an extension platform may have removable railings to facilitate the handling of materials.


## SAFETY PLANNING FOR HIGH CEILINGS AND WALLS

Stationary elevated work platforms, staging, and ladders used in interiors should comply with the requirements outlined in Chapter 15.

In using elevating work platforms to reach high ceilings and walls, the following requirements should be considered:


1. Elevating work platforms should be designed and constructed in accordance with the following ANSI standards: A992.3, "Manually Propelled Elevating Work Platforms"; A92.6, "Self-Propelled Elevating Work Platforms"; A92.5, "Boom-Supported Elevating Work Platforms."
2. Elevating work platforms should be operated, inspected, and maintained as specified in the operating manual for the equipment, and a record of inspections conducted while the unit is at the work site should be maintained at the work site.
3. Control levers on some elevating work platforms are unguarded and remain exposed to contact by overhead joists, trusses, beams, pipes, or other building parts when in a raised position. Hydraulic lifts sometimes drift or creep on their own, and this unintended movement can place control levers dangerously close to outside objects where they are vulnerable to activation.
4. Emergency controls accessible to people on the ground also are vulnerable to inadvertent activation by other objects and should be guarded.
5. Each individual control should have a different shape so that it can be easily identified when touched.
6. All boom-supported elevating work platforms should be equipped with an alarm or another suitable warning device at the platform. The alarm should be in operable condition and should automatically activate when the machine base is more than 5 degrees out of level in any direction.
7. The use of personal fall protection devices should be as specified in the manufacturer's operating manual; and if any personal fall protection devices are used, they must be secured only to manufacturer-approved hard points.
8. Only qualified personnel should be allowed to use elevating work platforms, by permit only, and they should be able to do the following:

a. Read and understand the manufacturer's operating manual, any associated rules and instructions, and all decals, warnings, and instructions on the platform.
b. Survey the work area for dropoffs or holes, bumps and floor obstructions, debris, overhead obstructions, elevated energy sources, and other possible hazards.
c. Ensure that the elevating work platform is on a firm, level surface.
d. Ensure that the work platform is loaded in accordance with the manufacturer's specifications.
e. Ensure that outriggers and/or stabilizers are used if required by the manufacturer.
f. Ensure that wheels are locked or chocked if the vehicle is on wheels.
g. Ensure that fall protection systems are in place.

## SAFETY PLANNING FOR PLASTERBOARD

Handling 4 × 8 sheets of plasterboard often is a difficult task because of their bulkiness and weight. Workers often are injured and boards broken because this 4 × 8 size is not suited for manual handling. New European Economic Community (EEC) standards call for using smaller sheets that are easier to handle than this size. The smaller sheets have resulted in reduced injury, labor costs, and breakage and increased productivity. Whenever a task is found to be injury-producing, as this one was by the EEC, the methods, procedures, and product design should be examined and safer ones substituted. Perhaps everyone should take a lesson from the EEC because it did something about safety, and its builders ended up with lower costs.

## SAFETY PLANNING FOR TOOLS USED IN INTERIOR FINISHING

Explosive-actuated tools probably lead the list of dangers associated with interior finishing, as nails and anchors can be propelled through a wall or board and speed onward like a bullet. The following guidelines should be considered:

1. Explosive-actuated (powder-actuated) tools should meet the design re-

quirements of ANSI A10.3, "Safety Requirements for Explosive-Actuated Tools."

2. Only qualified operators who have been authorized by permit should operate explosive-actuated tools, and they should:

   a. Be trained by an authorized instructor (one who has been trained, authorized, and provided an authorized instructor's card by the tool manufacturer or by an authorized representative of the tool manufacturer).

   b. Pass a written examination provided by the manufacturer of the tool.

   c. Possess a qualified operator's card supplied by the manufacturer and issued and signed by both the instructor and the operator.

3. Each tool should be provided with a lockable container with the words "POWDER-ACTUATED TOOL" in plain sight on the outside and a notice reading "WARNING—POWDER-ACTUATED TOOL TO BE USED ONLY BY A QUALIFIED OPERATOR AND KEPT UNDER LOCK AND KEY WHEN NOT IN USE" on the inside. Also, the operator's instruction and service manual, power load and fastener charts, the tool inspection record, and service tools and accessories should be kept in this container.

4. Daily inspection, cleaning, and testing should be performed as recommended by the manufacturer, with more extensive servicing given after each 1,000 fastenings.

5. Explosive-actuated tools should be tested, in accordance with the manufacturer's recommended procedure, each day before loading to see that safety devices are in the proper working condition.

6. Explosive-actuated tools and the charges should be secured at all times to prevent unauthorized possession or use.

7. Explosive-actuated tools should not be loaded until just prior to the intended fire time. Neither loaded nor empty tools are to be pointed at any employees, and workers' hands should be kept clear of the open barrel end.

8. The use of explosive-actuated tools must be prohibited in explosive or flammable atmospheres.

9. Fasteners should not be driven into soft or easily penetrable materials unless they are backed by a material that will prevent the fastener from passing: through to the other side; into very hard or brittle material, such as cast iron, hardened steel, glazed or hollow tile, glass block, brick, or rock; into concrete, unless the material thickness is at least three times the penetration of the fastener shank; or into spalled concrete.

## SAFETY PLANNING FOR WOODWORKING EQUIPMENT

1. All woodworking equipment should be inspected as it comes onto the job site, before use, as required in Chapter 14.
2. Circular saws should be equipped with guards that automatically and completely enclose the cutting edges, splitters, and anti-kickback devices.
3. All portable power-driven circular saws should be equipped with guards above and below the base plate or shoe. The upper and lower guards should cover the saw to the depth of the teeth, except for the minimum arc required to permit the base to be tilted for bevel cuts and the minimum arc required to allow proper retraction and contact with the work, respectively. When the tool is withdrawn from the work, the lower guard should automatically and instantly return to the covering position.
4. Blades of planers and jointers should be fully guarded and have cylindrical heads with throats in the cylinder.
5. Bandsaw blades should be fully enclosed except at the point of operation.
6. Automatic feeding devices should be installed on machines whenever possible. Feeder attachments should have the feed rolls or other moving parts covered or guarded to protect the operator from hazardous points.
7. The operating speed should be permanently marked on all circular saws over twenty inches in diameter or operating at over 10,000 peripheral feet per minute. Any saw so marked should not be operated at a speed other than that marked on the blade; and when a marked saw is retensioned for a different speed, the marking should be corrected to show the new speed.
8. Radial arm power saws should be equipped with an automatic brake.
9. The tables of radial arm or swing saws should extend beyond the leading edge of the saw blade.
10. Radial arm power saws should be installed in such a manner that the cutting head will return to the starting position when released by the operator. All swing cutoff and radial saws or similar machines that are drawn across a table should be equipped with limit stops to prevent the leading edge of the tool from traveling beyond the edge of the table.
11. Operators should use a push-stick, block, or other safe means on all operations close to high-speed cutting edges; must never use cracked, bent, or otherwise defective saw blades, cutters, knives, or other defective parts; and should not leave power saws running unattended.
12. A brush should be provided for the removal of sawdust, chips, and shavings on all woodworking machinery.

## SAFETY PLANNING FOR PAINTING

1. Packages containing paints, varnishes, lacquers, thinners, or other volatile painting materials should be kept tightly closed when not in use and should be stored in accordance with NFPA recommendations.
2. Paint materials in quantities other than the amount required for one day's use should not be stored in buildings under construction, and no more than a two-day supply should be stored in paint barges.
3. Unopened containers of paints, varnishes, lacquers, thinners, and other flammable paint materials should be kept in a well-ventilated location, free of excessive heat, smoke, sparks, flame, or direct rays of the sun.
4. Paint-soiled clothing and drop cloths, when not in use, should be stored in well-ventilated steel cabinets.
5. Paint scrapings and paint-saturated debris should be removed daily from the premises.
6. Ventilation adequate to prevent the accumulation of flammable vapors to hazardous levels should be provided in all areas where painting is done or paints are mixed.
7. No smoking, open flame, exposed heating elements, or other sources of ignition should be permitted in areas or rooms where spray painting is done.
8. Spray paint booths and equipment should be maintained in accordance with recommendations of the NFPA.
9. See Chapter 34 on fire prevention planning.

# Chapter *29*

## *Phase Planning for Exteriors*

Any inanimate object, regardless of its position or configuration, may be
expected to perform at any time in a totally unexpected manner for reasons
that are either entirely obscure or else completely mysterious.

(Flap's law)

Phase planning for exteriors will consider two basic hazards:

1. *Powerline contact:* Powerlines sometimes are installed too close to the
   building. Table 234-1 in *The National Electrical Safety Code* (NESC)
   allows for uninsulated, high-voltage conductors to be as close as five feet
   in horizontal distance to buildings, and in some circumstances as close as
   three feet. With such close horizontal clearance, the installation of metal
   sheathing and the use of rolling scaffolds, aerial lifts, suspended scaffolds,
   and ladders are quite hazardous. When hazards overlooked at the time of
   design are discovered during phase planning, appropriate change orders
   must be issued to eliminate the hazard. In the case of powerlines too close
   to the work area, change orders should include requesting the electric
   utility to bury or relocate them. Seeing that powerlines are inaccessible
   also proves to be valuable for future painting, window washing, or other
   exterior maintenance.
2. *Falls from an elevation:* This is the highest-risk hazard that will be
   encountered by those installing, finishing, and maintaining outside sur-

faces. The information on fall prevention in Chapter 15 will be most helpful in avoiding this hazard.

Access to building exteriors usually is provided by various forms of suspended scaffolding.

## SAFETY PLANNING TO ENSURE SAFE SUSPENDED SCAFFOLDS

1. All parts of all suspended scaffolds, except support cables, should have a minimum safety factor of four.
2. Suspended scaffolds should be supported by ropes secured to outrigger beams. A minimum safety factor of six is required for suspension ropes.
3. Suspension ropes should have the fixed end equipped with a proper-size thimble secured by splicing or other means, and be attached to the support by a closed shackle. Running ends should be attached to the hoisting drum, and at least four turns of rope should remain on the drum at all times.
4. Suspension ropes should be attached at the vertical centerline of the outrigger, and the attachment should be directly over the hoisting drum.
5. Suspended scaffolds should be guyed, braced, or equipped with a tagline to prevent swaying.
6. Stone setters and two-point suspension scaffolds should not be joined by bridging.
7. Brackets should be of wrought iron or mild steel. No reinforcing steel should be used as any part of a support system. Brackets should have attachments for guardrails, intermediate rails, and toeboards. Tiebacks of ¾-inch manila rope, or the equivalent, should serve as a secondary means of anchorage, installed at right angles to the building, whenever possible, and secured to a structurally sound portion of the building.
8. Suspended scaffolds should have a standard railing.
9. Overhead protection equal to ¾-inch plywood should be provided on a multiple-point suspension mason's scaffold.
10. Scaffold machines, either electromechanically powered or hand-powered, should be worm-geared or powered both ways (up or down). The design must be such that when the power is stopped, the scaffold cannot move. Hoisting machines should be of a type tested and listed by UL or FM.
11. Controls for powered scaffolds should be of the deadman type with a nonlocking switch.

12. A device to shut off the power should be installed ahead of the operating control.
13. Hand-operated release mechanisms that can permit the load to descend faster than the speed rating should be prohibited.
14. Scaffold machines with cast metal parts should be prohibited.
15. Every suspended scaffold should be tested with twice the maximum loading before being put into operation.
16. All anchorages of a suspended scaffold should be inspected at the beginning of each shift.
17. Each person supported by a suspended scaffold should be protected by an approved safety belt attached to a lifeline. The lifeline should be securely attached to substantial members of the structure (not the scaffold), or to securely rigged lines, which will safety suspend the employee in case of a fall. To keep the lifeline continuously attached, with a minimum of slack, to a fixed structure, the attachment point of the lifeline should be changed as the work progresses. An exception occurs for lower tiers of a multistage suspension scaffold or where overhead protection is required. In these situations the lifeline should be securely attached to the scaffold, and independent wire ropes should be used at each end of the suspended scaffold, with approved grabbing and locking devices. (See Figure 29-1.)
18. A mason's adjustable multiple-point suspension scaffold should be capable of sustaining a load of fifty pounds per square foot and should not be overloaded.
19. The mason's scaffold should be provided with hoisting machines of a type tested and listed by UL or FM.
20. The scaffold outrigger beams for the mason's adjustable multiple-point suspension scaffold should consist of structural metal securely fastened or anchored to the frame or the floor system of the structure. Each outrigger beam should be equivalent in strength to at least a standard seven-inch, 15.3-pound steel I-beam, at least fifteen feet long, and should not project more than six feet, six inches beyond the bearing point. Where the overhang exceeds six feet, six inches, outrigger beams should be composed of stronger beams or multiple beams and be installed under the supervision of a competent person.
21. All outrigger beams should be set and maintained with their webs in a vertical position and anchored to the structure by the equivalent of U-bolts and anchor plates, washers, and nuts.
22. A stop bolt should be placed at each end of every outrigger beam.
23. The outrigger beam should rest on wood-bearing blocks.
24. The free end of the suspension wire ropes should be equipped with

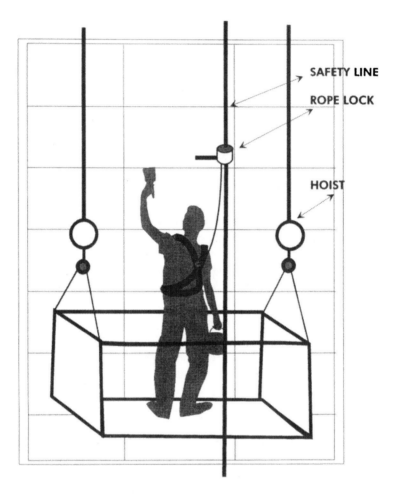

SAFETY LINE

ROPE LOCK

HOIST

**Figure 29-1.**  Suspended scaffold safety lines.

proper-size thimbles and secured by splicing or other equivalent means.
The running ends should be securely attached to the hoisting drum, and at
least four turns of rope should at all times remain on the drum.

25.  Where a single outrigger beam is used, the steel shackles or the clevises
with which the wire ropes are attached to the outrigger beams should be
placed directly over the hosting drums.

26.  When employees are working on a scaffold where an overhead hazard
exists, overhead protection consisting of two-inch planking or material of
equivalent strength, laid tight and extending not less than the width of the

scaffold, should be provided not more than nine feet above the scaffold planking.

27. Each scaffold should be installed or relocated under the supervision of a competent person.

28. Stone setters' adjustable multiple-point suspension scaffolds should be capable of sustaining a working load of twenty-five pounds per square foot, and should not be overloaded. Scaffolds should not be used for storage of stone or other heavy materials.

29. The hoisting machines and supports used with stonecutters' scaffolds should be of a type tested and listed by UL or FM.

30. The platform for the stone setters' scaffolds should be securely fastened to the hangers by U-bolts or other equivalent means.

31. The stonecutters' scaffolds should be suspended from metal outriggers, iron brackets, wire rope slings, or iron hooks.

32. When two or more stonecutters' scaffolds are used on a structure, they should not be bridged one to the other, but should be maintained at an even height with the platforms abutting closely.

33. Two-point suspension scaffold platforms should be not less than twenty inches or more than thirty-six inches wide. The platform should be fastened to the hangers by U-bolts or by other equivalent means.

34. The hangers of two-point suspension scaffolds should be made of mild steel, or equivalent materials, having a cross-sectional area capable of sustaining four times the maximum rated load, and should be designed with a support for a standard railing.

35. The hoisting equipment used on two-point suspension scaffolds should be of a type tested and listed by UL or FM.

36. The roof irons or hooks should be made of mild steel or another equivalent material of proper size and design, securely installed and anchored. Tie-backs of ¾-inch manila rope or the equivalent should serve as a secondary means of anchorage, installed at right angles to the face of the building, whenever possible, and secured to a structurally sound portion of the building.

37. Two-point suspension scaffolds should be suspended by wire, synthetic, or fiber ropes capable of supporting at least six times the rated load. All other components should be capable of supporting at least four times the rated load.

38. The sheaves of all blocks, consisting of at least one double and one single block, should fit the size and the type of rope used.

39. All wire ropes, fiber and synthetic ropes, slings, hangers, platforms, and other supporting parts should be inspected before every installation. Periodic inspection should be made while the scaffold is in use.

40. On suspension scaffolds designed for a working load of 500 pounds, no more than two employees should be permitted to work at one time. On suspension scaffolds with a working load of 750 pounds, no more than three people should be permitted to work at one time. Each employee should be protected by an approved safety belt attached to a lifeline. The lifeline should be securely attached to substantial members of the structure (not the scaffold) or to securely rigged lines, which will safely suspend the employee in case of a fall. To keep the lifeline continuously attached to a fixed structure with a minimum of slack, the attachment point of the lifeline should be appropriately changed as the work progresses.

41. Two-point suspension scaffolds should be securely lashed to the structure. Window cleaners' anchors should not be used.

## SAFETY PLANNING FOR CRANE-SUPPORTED SCAFFOLDING

1. Crane-supported scaffolds may be used only under strictly controlled conditions and when authorized by permit issued by the construction manager.
2. Crane-supported scaffolding should conform to the following:
   a. The scaffold should be of metal or metal frame construction with a forty-two-inch handrail, an intermediate rail, and four-inch-minimum toeboards.
   b. The hook connection to the scaffold rigging should be closed by a positive clamping device that cannot open because of the load position on the hook. A swivel placed between the hoist line and the hook will prevent line rotation.
   c. The scaffold should be stabilized at all times by guylines or taglines.
   d. Only cranes with power-operated up-and-down boom hoists and load lines should be used to support these scaffolds. Scaffolds should be lowered only under power and not by the brake.
   e. Thorough inspection by a qualified person should be made of all hoisting parts prior to each use.
   f. A competent supervisor should observe the operations while personnel are working from crane-supported scaffolds.
   g. Lifting bridles should consist of four legs attached so that platform stability is assured.
   h. All scaffold components should be capable of supporting at least four times the maximum intended load.

3. All crane-supported personnel baskets and scaffolding should meet these additional requirements taken from OSHA §1926.550, "Cranes and Derricks":

   a. The use of a crane or derrick to hoist employees on a personnel platform is prohibited, except when the erection, use, and dismantling of conventional means of reaching the work site, such as a personnel hoist, ladder, stairway, aerial lift, elevating work platform or scaffold, would be more hazardous, or is not possible because of the structural design or work-site conditions.

   b. Hoisting of the personnel platform should be performed in a slow, controlled, cautious manner with no sudden movements of the crane or the derrick, or of the platform.

   c. Load lines should be capable of supporting, without failure, at least seven times the maximum intended load, except that where rotation-resistant rope is used, the lines should be capable of supporting, without failure, at least ten times the maximum intended load. The required design factor is achieved by taking the current safety factor of 3.5 (required under §1926.550(b)(2)) and applying the 50 percent derating of the crane capacity that is required by §1926.550(g)(3)(i)(F).

   d. Load and boom hoist drum brakes, swing brakes, and locking devices such as pauls or dogs should be engaged when the occupied personnel platform is in a stationary working position.

   e. The crane should be uniformly level within one percent of level grade and located on firm footing. Cranes equipped with outriggers should have them fully deployed following the manufacturer's specifications, insofar as applicable, when hoisting employees.

   f. The total weight of the loaded personnel platform and related rigging should not exceed 50 percent of the rated capacity for the radius and the configuration of the crane or the derrick.

   g. The use of machines having live booms (booms in which lowering is controlled by a brake without aid from other devices that slow the lowering speeds) is prohibited.

   h. Cranes and derricks with variable-angle booms should be equipped with a boom angle indicator, readily visible to the operator.

   i. Cranes with telescoping booms should be equipped with a device to indicate clearly to the operator, at all times, the boom's extended length, or an accurate determination of the load radius to be used during the lift should be made before hoisting of personnel.

   j. A positive-acting device should be used that prevents contact between the load block or the overhaul ball and the boom tip (anti-two-blocking device), or a system should be used that deactivates the hoisting action before damage occurs in the event of a two-blocking situation (two-block damage prevention feature).

k. The load line hoist drum should have a system or a device on the power train, other than the load hoist brake, that regulates lowering of the rate of speed of the hoist mechanism (controlled load lowering). Free fall is prohibited.

## SAFETY PLANNING FOR PERSONNEL PLATFORMS

The following guidelines have been taken from OSHA §1926.550, "Cranes and Derricks":

1. Personnel platform design criteria: The personnel platform and suspension system should be designed by a qualified engineer or a qualified person competent in structural design.
2. The suspension system should be designed to minimize tipping of the platform due to movement of employees occupying it.
3. The personnel platform itself, except the guardrail system and body belt/harness anchorages, should be capable of supporting, without failure, its own weight and at least five times the maximum intended load. Criteria for guardrail systems and body belt/harness anchorages are contained in Subparts E and M of the OSHA document.
4. Platform specifications: Each personnel platform should be equipped with a guardrail system that meets the requirements of Subpart M of the OSHA document and should be enclosed at least from the toeboard to midrail with either solid construction or expanded metal having openings no greater than $\frac{1}{2}$ inch (1.27 cm).
5. A grab rail should be installed inside the entire perimeter of the personnel platform.
6. Access gates, if installed, should not swing outward during hoisting.
7. Access gates, including sliding or folding gates, should be equipped with a restraining device to prevent accidental opening.
8. Headroom should be provided that allows employees to stand upright in the platform.
9. In addition to the use of hard hats, employees should be protected by overhead protection on the personnel platform when they are exposed to falling objects.

10. All rough edges exposed to contact by employees should be surfaced or smoothed in order to prevent injury to employees from punctures or lacerations.

11. All welding of the personnel platform and its components should be performed by a qualified welder familiar with the weld grades, types, and material specified in the platform design.

12. The personnel platform should be conspicuously posted with a plate or another permanent marking that indicates the weight of the platform and its rated load capacity or maximum intended load.

13. Personnel platform loading: The personnel platform should not be loaded in excess of its rated load capacity. When a personnel platform does not have a rated load capacity, then it should not be loaded in excess of its maximum intended load.

14. The number of employees occupying the personnel platform should not exceed the number required for the work being performed.

15. Personnel platforms should be used only for employees, their tools, and the materials necessary to do their work, and should not be used to hoist only materials or tools when not hoisting personnel.

16. Materials and tools intended for use during a personnel lift should be secured to prevent displacement.

17. Materials and tools intended for use during a personnel lift should be evenly distributed within the confines of the platform while the platform is suspended.


SAFETY PLANNING FOR RIGGING WORK PLATFORMS

OSHA §1926.550, "Cranes and Derricks," further requires:

1. When a wire rope bridle is used to connect the personnel platform to the load line, each bridle leg should be connected to a master link or a shackle in such a way that the load is evenly divided among the bridle legs.

2. Hooks on overhaul ball assemblies, lower load blocks, or other attachment assemblies should be of a type that can be closed and locked, eliminating the hook throat opening. Alternatively, an alloy anchor-type shackle with a bolt, nut, and retaining pin may be used.

3. Wire rope, shackles, rings, master links, and other rigging hardware must be capable of supporting, without failure, at least five times the maximum intended load applied or transmitted to that component. Where rotation-

resistant rope is used, the slings should be capable of supporting without failure at least ten times the maximum intended load.

4. All eyes in wire rope slings should be fabricated with thimbles.

5. Bridles and associated rigging for attaching the personnel platform to the hoist line should be used only for the platform and the necessary employees, their tools, and the materials necessary to do their work, and should not be used for any other purpose when not hoisting personnel.

## SAFETY PLANNING FOR TRIAL LIFTS, INSPECTION, AND PROOF-TESTING

The following also are OSHA requirements:

1. A trial lift with the unoccupied personnel platform loaded at least to the anticipated liftweight should be made from ground level, or any other location where employees will enter the platform, to each location at which the personnel platform is to be hoisted and positioned. This trial lift should be performed immediately before placing of personnel on the platform. The operator should determine that all systems, controls, and safety devices are activated and functioning properly; that no interferences exist; and that all configurations necessary to reach those work locations will allow the operator to remain under the 50 percent limit of the hoist's rated capacity. Materials and tools to be used during the actual lift can be loaded in the platform, as provided in paragraphs (g)(4)(iii)(D) and (E) of the OSHA document for the trial lift. A single trial lift may be performed at one time for all locations that are to be reached from a single setup position.

2. The trial lift should be repeated prior to hoisting of employees whenever the crane or the derrick is moved and set up in a new location or returned to a previously used location. Additionally, the trial lift should be repeated when the lift route is changed unless the operator determines that the route change is not significant (i.e., the route change would not affect the safety of hoisted employees).

3. After the trial lift, and just prior to the hoisting of personnel, the platform should be hoisted a few inches and inspected to ensure that it is secure and properly balanced. Employees should not be hoisted unless the following conditions are determined to exist:

   a. Hoist ropes are free of kinks.

    b.  Multiple part lines are not twisted around each other.

    c.  The primary attachment is centered over the platform.

    d.  The hoisting system has been inspected if the load rope was slack to ensure that all ropes are properly seated on drums and in sheaves.

4.  A visual inspection of the crane or derrick, rigging, personnel platform, and crane or derrick base support or ground should be conducted by a competent person immediately after the trial lift to determine whether the testing has exposed any defect or produced any adverse effect upon any component or structure.

5.  Any defects found during inspections that create a safety hazard should be corrected before the hoisting of personnel.

6.  At each job site, prior to the hoisting of employees on the personnel platform, and after any repair or modification, the platform and the rigging should be proof-tested to 125 percent of the platform's rated capacity by holding it in a suspended position for five minutes with the test load evenly distributed on it (this may be done concurrently with the trial lift). After proof-testing, a competent person should inspect the platform and the rigging. Any deficiencies found should be corrected, and another proof-test should be conducted. Personnel hoisting should not be conducted until the proof-testing requirements are satisfied.

7.  Work practices: Employees should keep all parts of their bodies inside the platform during raising, lowering, and positioning. This provision does not apply to an occupant of the platform performing the duties of a signal person.

8.  Before employees exit or enter a hoisted personnel platform that is not landed, the platform should be secured to the structure where the work is to be performed, unless securing to the structure creates an unsafe situation.

9.  Taglines should be used unless their use creates an unsafe condition.

10. The crane or derrick operator should remain at the controls at all times when the crane engine is running and the platform is occupied.

11. Hoisting of employees should be promptly discontinued upon indication of any dangerous weather conditions or other impending danger.

12. Employees being hoisted should remain in continuous sight of and in direct communication with the operator or the signal person. In those situations where direct visual contact with the operator is not possible, and the use of a signal person would be hazardous for that person, direct communication alone such as by radio may be used.

13. Except over water, employees occupying the personnel platform should use a body belt/harness system with lanyard appropriately attached to the lower load block or overhaul ball, or to a structural member within the

personnel platform capable of supporting a fall impact for employees using the anchorage. In working over water, the requirements of §1926.106 should apply.

14. No lifts should be made on another of the crane's or the derrick's load lines while personnel are suspended on a platform.

15. Hoisting of employees while the crane is traveling is prohibited, except for portal, tower, and locomotive cranes, or when the employer demonstrates that there is no less hazardous way to perform the work.

16. Under any circumstances where a crane would travel while hoisting personnel, the employer should implement the following procedures to safeguard employees:

    a. Crane travel should travel be restricted to a fixed track or runway.

    b. Travel should be limited to the load radius of the boom used during the lift.

    c. The boom must be parallel to the direction of travel.

    d. A complete trial run should be performed to test the route of travel before employees are allowed to occupy the platform. This trial run can be performed at the same time as the trial lift that tests the route of the lift.

    e. If travel is done with a rubber-tired carrier, the condition and the air pressure of the tires should be checked. The chart capacity for lifts on rubber should be used for application of the 50 percent reduction of rated capacity. Notwithstanding paragraph (g)(3)(i)(E) of the OSHA document, outriggers may be partially retracted as necessary for travel.

17. A meeting attended by the crane or derrick operator, signal person(s) (if necessary for the lift), employee(s) to be lifted, and person responsible for the task to be performed should be held to review the appropriate requirements of paragraph (g) of the OSHA document and the procedures to be followed. This meeting should be held prior to the trial lift at each new work location, and should be repeated for any employees newly assigned to the operation.


## SAFETY PLANNING FOR MASONRY WALLS ERECTED FROM THE GROUND UP

1. A limited-access zone should be established whenever a masonry wall is being constructed, and such zone should be:

    a. Established prior to the start of construction on the wall.

    b. Equal to the height of the wall to be constructed plus four feet and run the entire length of the wall.

    c. Established on the side of the wall that will be unscaffolded.

    d. Restricted to entry by employees actively engaged in constructing the wall; no other employees should be permitted to enter the zone.

    e. Kept in place until the wall is adequately supported to prevent overturning and to prevent collapse unless the height of the wall is over eight feet, in which case the limited-access zone should remain in place until braced.

2. All masonry walls over eight feet in height should be adequately braced to prevent overturning and to prevent collapse unless the wall is adequately supported so that it will not overturn or collapse. The bracing should remain in place until permanent supporting elements of the structure are in place.

3. Scaffolds for masonry construction workers should not be used to provide temporary lateral support of masonry walls.

4. Clean-outs should be on the side of the masonry wall opposite the scaffolding.

5. Some masonry walls must be erected along narrow alleys or roadways parallel to powerlines. The use of scaffolding to build the wall farther restricts the space between the wall and the powerlines. In no circumstances should a boom truck delivering pallets of block or brick be allowed to enter this confined space to unload pallets onto the scaffolding or at designated locations beside the scaffolding because the boom cannot be safely rotated 360 degrees fully extended without risking contact with the powerlines. The only safe way to deliver pallets to such a location is to offload the pallets from the truck where the boom cannot reach the powerlines and then move them by forklift to the wall that is being built.

# Chapter *30*

## *Phase Planning for Roofs*

Few things are brought to a successful issue by impetuous desire, but most
by calm and prudent forethought.

(Thucydides)

**M**any workers are injured or killed by falling off a roof; so phase planning must include safeguards so that roof construction and maintenance can
be done safely. The safest access to roofs is through the interior of the
building. (See Chapter 15 on fall prevention for additional information.) If
the roof has a parapet or a guardrail around its perimeter, falls from an
elevation usually are significantly reduced. A steeply pitched roof is a hazard
in itself and is expensive to maintain because of the difficult working surface.
Roofers in home building construction have the highest incidence of falls
from an elevation. The requirements for fall protection now have been
greatly expanded by OSHA §1926.500, Subpart M.[1,2]

Prefabricated metal roof panels have been known to collapse when used as
a temporary support in installing insulation. In one instance, a worker fell
approximately sixty feet to his death when helping to install insulation and using
a roof panel for temporary access. This panel was about two feet wide and
sixteen feet long, with upright edges approximately four inches high along both
sides of the panel's length. Apparently the worker's knee or foot pushed one
of the upright edges outward, causing the panel to buckle near the center and
collapse. These roofing panels appear to give safe support but should not be

used for temporary access unless two pieces of 2 × 10 planking the same length as the panel are laid flat on it. A warning about this hazard is now printed on some roof panels and the use of planking advised. The use of body belts will be prohibited after January 1, 1998, and harnesses will be required after that date. (Figure 30-1 shows safety lines for roofs.)

Phase planning should include a requirement that erection methods be examined by a registered professional structural engineer to determine the ability of the roof to support personnel and materials while it is being constructed. One common cause of roof failure during construction is the placing of building materials on joists set on edge but not yet braced. The slightest side movement can cause the joists to tip over sideways in a domino effect, dumping the materials and the workers onto anyone working below.

## SAFETY PLANNING FOR SAFE ROOFING DEVICES AND PRACTICES

1.  In the construction, demolition, repair, or other use of roofs, protective devices should be provided that will prevent personnel from slipping and falling from the roof and prevent personnel on lower levels from being struck by falling objects.
2.  Workers engaged in the construction, maintenance, or repair of built-up roofing (but not construction of the roof deck) on roofs having a slope less than or equal to four vertical to twelve horizontal and with a ground-to-eave height greater than sixteen feet should be protected by either a motion-stopping-safety system (MSS) or a warning line system.
3.  Protective devices may be shingling footlocks, lifelines, crawling boards, ladders, railings, catwalks, toeboards, temporary decking, or any other device that will furnish protection.
4.  On all roofs more than sixteen feet in height, a hoisting device, stairways, or progressive platforms should be furnished for supplying materials and equipment.
5.  Level, guarded platforms should be provided at the landing area on the roof.
6.  Crawling boards should be not less than ten inches wide and one inch thick, having 1″ × 1½″ cleats. The cleats should be equal in length to the width of the board and spaced at equal intervals not to exceed twenty-four inches. Nails should be driven through and clinched on the underside. The crawl-

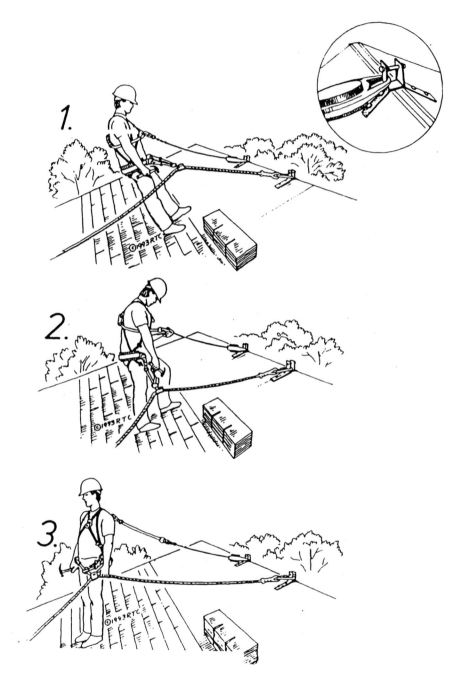

**Figure 30-1.** Safety lines for roofs. (Courtesy of the Research and Trading Corporation.)

ing board should be secured and should extend from the ridge pole to the eaves when used with roof construction, repairs, or maintenance.

7. A firmly fastened lifeline of at least ¾-inch-diameter rope, or the equivalent, should be strung beside each crawling board for a handhold.

8. Roofing brackets should be secured by nailing, in addition to the pointed metal projections. When it is impractical to nail brackets, rope supports should be used. When rope supports are used, they should consist of first-grade ¾-inch-diameter manila rope, or the equivalent.

9. Roofing materials and accessories that are on the roof and unattached and could be moved by the wind, including metal roofing panels, should be secured when wind speeds are greater than, or are anticipated to exceed, ten miles per hour.

10. When warning line systems are required by the construction manager, they should comply with the following guidelines:

    a. Warning lines should be erected around all sides of the work area.

       (1) When mechanical equipment is not being used, the warning line should be erected not less than six feet from the roof edge.

       (2) When mechanical equipment is being used, the warning line should be erected not less than six feet from the roof edge that is parallel to the direction of mechanical equipment operation and not less than ten feet from the roof edge that is perpendicular to the direction of mechanical equipment operation.

    b. Warning lines should consist of a rope, wire, or chain and supporting stanchions erected as follows:

       (1) The rope, wire, or chain shall be flagged at at least six-foot intervals with high-visibility material.

       (2) The rope, chain, or wire should be rigged and supported in such a way that its lowest point (including sag) is no less than thirty-four inches from the roof surface and its highest point no more than thirty-nine inches from the roof surface.

       (3) After being erected with the rope, wire, or chain attached, stanchions should be capable of resisting, without tipping over, a force of at least sixteen pounds applied horizontally against the stanchion thirty inches above the roof surface, perpendicular to the warning line, and in the direction of the roof edge.

       (4) The rope, wire, or chain should have a minimum tensile strength of 300 pounds, and after being attached to the stanchions should be capable of supporting, without breaking, the loads applied to the stanchions [as described in item (3) above].

       (5) The lines should be attached at each stanchion in such a way that

pulling on one section of the line between stanchions will not result in slack being taken in adjacent sections before the stanchion tips over.

   c. Access paths should be erected as follows:

      (1) Points of access, materials-handling areas, and storage areas should be connected to the work area by a clear access path formed by two warning lines.

      (2) When the path to a point of access is not in use, a rope, wire, or chain, equal in strength and height to the warning line, should be placed across the path at the point where the path intersects the warning line erected around the work area.

11. Those working in a roof-edge materials-handling or storage area located on a roof having a slope less than or equal to four vertical to twelve horizontal and with a ground-to-eave height greater than sixteen feet should be protected by the use of an MSS system along all unprotected roof sides and edges of the area.

   a. When guardrails are used at hoisting areas, a minimum of four feet of guardrail should be erected on each side of the access point through which materials are hoisted.

   b. A chain or a gate should be placed across the opening between the guardrail sections when hoisting operations are not taking place.

   c. When guardrails are used at bitumen pipe outlets, a minimum of four feet of guardrail should be erected on each side of the pipe.

   d. When safety belt systems are used, they should not be attached to the hoist.

   e. When safety belt systems are used, they should be rigged to allow the movement of workers only as far as the roof edge.

   f. Materials should not be stored within six feet of the roof edge unless guardrails are erected at the roof edge.

   g. Materials that are to be piled, stacked, or grouped should be stable and self-supporting.

## Safety Planning for Use of Hot Roofing Materials

1. Heating devices or melting kettles should be placed on a level, firm foundation and protected against traffic, accidental tipping, or similar hazard.

2. Enclosed areas in which hot substances are being heated or applied should be ventilated.

3. Fire extinguishers, rated not less than 10-B, should be available at all locations where heating devices or melting kettles are in use.
4. Heating devices or melting kettles when in use should not be left unattended.
5. Bituminous-material melting kettles should be provided with an effective lid or hood and a thermometer in operating condition.
6. Bituminous-material melting kettles should not be used or operated inside of or on buildings, or within twenty-five feet of buildings or combustible materials.
7. Ladles, equipment, and material should be moisture-free before being used or placed in heated material.
8. Runways or passageways, clear of obstructions, should be provided for all persons carrying hot substances.
9. Hot substances should not be carried up or down ladders.
10. Hoisting gear used in handling hot substances should be adequate for loads imposed and should be securely braced and anchored.
11. Protection should be provided against contact, radiant heat, glare, fumes, and vapors for all persons handling hot substances.
12. Buckets or vessels for handling of hot substances should be substantially constructed and free from any soldered joints or attachments.
13. Containers for handling and transportation of hot substances should not be filled higher than four inches from the top.
14. See Chapter 34, on fire prevention planning, for additional information.

## REFERENCES

1. U.S. Department of Labor, OSHA, "Safety Standards for Fall Protection in the Construction Industry: Final Rule," CFR Parts 1910 and 1926, *Federal Register,* Part III, Tuesday, Aug. 9, 1994.
2. Ellis, J. Nigel, *Fall Protection,* 2nd Edition, American Society of Safety Engineers, Des Plaines, IL, 1994.

# Chapter 31

## *Phase Planning for Maintenance*

> When a stupid man is doing something he is ashamed of, he always declares
> it is his duty.
>
> (George Bernard Shaw)

Maintenance and repair of construction equipment sometimes must be done under adverse conditions—in temporary facilities, in the field, under adverse weather or other hazardous conditions such as the presence of toxic substances, flammable liquids, or roadway traffic in the immediate vicinity, and without proper tools or equipment for specific repairs such as adequate lifting equipment and substantial blocking or cribbing for work under equipment; and these activities can cause injury and death if proper attention is not given them in phase planning and alternate procedures are not devised.

Toxic substances often are used as cleaning agents. OSHA §1926.59, "Hazard Communication," provides detailed information on hazard communication. Subparagraph (g), covers manufacturers' material safety data sheets (MSDS). All cleaning materials should be itemized in the phase plan, and material safety data sheets should be provided for every hazardous substance used in maintenance. (See Chapter 16 for further details.)

### SAFETY PLANNING FOR USE OF HEATING DEVICES AND SYSTEMS

The use of portable unvented heaters in maintenance shops in cold weather must not be allowed. They eat up the oxygen in the air and produce carbon monoxide if proper ventilation is not provided, and have been responsible for the deaths of many people. The following are good guidelines for temporary or permanent heating systems for maintenance shops.

1. Where smoke pipes from heating apparatus pass through combustible walls or roofs, UL- or FM-approved properly insulated and ventilated sleeves or roof jacks should be used.
2. All smoke pipe joints should be securely fastened and supported to prevent accidental displacement or separation.
3. Spark arresters should be provided on all smokestacks or burning devices that have forced drafts or short stacks permitting live sparks or hot materials to escape.
4. Only temporary heating devices approved by the designated authority should be used. A positive operating procedure should be established to assure the following controls:
   a. Proper placement and servicing.
   b. Safe clearance from combustible material.
   c. Close surveillance.
   d. Safe fuel storage and refueling.
   e. Proper maintenance.
   f. Determination of gaseous contamination or oxygen deficiency.
5. Open-flame heating devices having exposed fuel below the flame are prohibited.
6. Gas or liquid-fueled space-heating devices used in any enclosed building, room, or structure should be vented by the flue pipe to the exterior of the structure. UL- or FM-listed heating devices may be used if they are vented to the outside.
7. Heating equipment may be installed in lubrication or service areas where there is no dispensing or transfer of flammable liquids, provided that the bottom of the heating unit is at least eighteen inches above the floor and is protected from damage.
8. Heating equipment installed in lubrication or service areas where flammable liquids are dispensed should be of a type approved for garages, and should be installed at least eight feet above the floor.
9. Flammable liquid-fired heaters should be equipped with a primary safety control to stop the flow of fuel in the event of flame failure. Barometric or gravity oil feed is not a primary safety control.
10. Heaters used near combustible tarpaulins, canvas, or similar coverings should be located at least ten feet from such combustibles and should be securely fastened to prevent ignition or upsetting of the heater.

## SAFETY PLANNING FOR SAFE LOCKOUT AND TAGGING OF CIRCUITS

Equipment should be locked out when emergency field repairs are made to avoid having someone other than the repair person start it up unexpectedly. Because rock-crushing plants have many conveyors, they pose great danger to the person making repairs if another person can inadvertently turn the equipment back on. Be sure that any construction equipment in for repair is locked out so that someone cannot use it just because it is conveniently located. A worker was killed when a mechanic used a crane that was sent in to have an anti-two-blocking device installed and was not locked out; the mechanic two-blocked it and dropped the hook on the helper. Lockout systems are needed for both electrical and mechanical systems, using the following guidelines:

1. Before workers perform any servicing or maintenance on a system where unexpected energizing, start-up, or release of kinetic or stored energy could occur and cause injury or damage, the system should be isolated. Workers and resources should not be considered protected until hazardous energy control procedures have been implemented.
2. All control activities should be coordinated with and approved by the construction manager. When contractors are planning the use of hazardous energy control procedures, they should submit their hazardous energy control plan to the construction manager for acceptance prior to initiation of these activities. The construction manager should coordinate the activities with the various contractors throughout the planning and the implementation of these activities. They each should inform the others of their energy control procedures, ensure that their own personnel understand and comply with the rules and restrictions of the procedures, and ensure that their employees affected by the hazardous energy control activity are notified when the procedural steps outlined in the hazardous energy control plan are to be initiated.
3. When energy control procedures affect existing operations, a preparatory inspection by the phase supervisor and contractor personnel should be conducted to ensure that all affected employees understand the energy hazards and the procedures for their control. The preparatory meeting should be documented. The time and date of the meeting, the subject matter discussed, and the names of all employees attending should be recorded.
4. Lockout and tagout should be performed only by authorized workers designated by permit.

5. All workers affected by lockout or tagout should be notified prior to and upon completion of the attachment and the removal of lockout or tagout devices.

6. Systems with energy isolating devices that are capable of being locked out should utilize locking devices to control hazardous energy. If an energy isolating device is not capable of being locked out, the hazardous energy control procedures should require that the tagout device be attached to the same location, if possible, where the lockout device would have been attached. If this is not possible, then the tag should be attached as close as is safely possible to the device and in a position that will be immediately obvious to anyone attempting to operate the device. If it is appropriate, the entire danger area should be fenced off or barricaded. Alternatives to consider if a lockout device cannot be used include: the tag is placed so that the energy isolating device cannot be operated; the isolating circuit mechanism can be removed; the control switch can be blocked; an extra disconnecting device can be opened; a valve handle to reduce the likelihood of inadvertent energization can be removed.

7. Lockout and tagout devices should:
   a. Be used individually, with workers not relying on each other's devices.
   b. Withstand the abuse to which they may be exposed (weather, corrosion) so that their message does not become illegible.
   c. Identify who put them on.
   d. Be difficult to remove.
   e. Have a standardized format.
   f. Have tags and seals that are not reusable.
   g. Be attachable by hand, self-locking, and nonreleasable.
   g. Identify the hazardous condition, what can happen, and what to do (in bold letters).
   h. Tell what is shut down and for how long.

8. Phase supervisors should monitor and inspect equipment to see that workers are using the lockout and tagout systems properly.

## SAFE PLANNING OF EQUIPMENT REPAIRS

1. Workers should follow manufacturers' recommended maintenance procedures.

2. All machinery or equipment should be shut down with positive means taken to prevent its operation while repair or manual lubrication is being done.

3. Workers should not repair equipment in traffic areas. If repairs must be made on roads, a yellow flashing light or four-way flashers visible from all directions should be used.
4. It is necessary to block or crib heavy machinery, equipment, or parts suspended or held apart by slings, hoists, or jacks before workers are allowed underneath or between them.
5. Bulldozer and scaper blades, end-loader buckets, dump bodies, and similar equipment should be fully lowered or blocked when being repaired or when not in use. All controls should be in a neutral position, with engines stopped and brakes set, unless work that is being performed requires them to be otherwise.
6. Latticework crane booms should be lowered onto blocking or cribbing to support each section before it is disassembled.
7. All mobile equipment and the areas in which it is operated should have adequate illumination.
8. Towing devices should be structurally adequate for the weight drawn, and should be securely mounted. No one should be allowed to get between a towing vehicle and the piece being towed until both have been completely stopped with all brakes set and wheels chocked.

# Chapter 32

## Economic Incentives for Construction Safety Planning

. . . if injuries are to be controlled by the legislative system, not only must the legislative requirements be relevant to the true causes of injury, but there must also be compliance with those requirements. The implications of this are simple. Compliance implies enforcement and enforcement requires a police function and a set of appropriate sanctions. In industry, policing and sanction-applying are expensive, nonproductive and time-consuming activities which divert effort that could otherwise be devoted to alternative strategies.

(E. C. Wigglesworth)

Now that the major steps in construction safety planning have been discussed, the subject of making planning cost-effective needs to be addressed.

We all have witnessed the collapse of Communism and the necessity for a market economy. In our humanitarian, democratic society, the value of human life has slowly been taking precedence over profit, as people are no longer considered an expendable resource. This perspective has made the cost of injuries so high that entrepreneurs now recognize that hazard prevention is a viable alternate profit strategy. This is creating a market economy for safety, where the role of the three branches of government—legislative, executive, and judicial—should not be to create monstrous policing and enforcement agencies but to have the common goal of collecting and providing hazard information.

As this market economy for safety develops, it will enhance the earning power of hazard prevention, which has never been fully appreciated. Construction's most pressing and immediate need is to have a source of pertinent hazard information that can be incorporated into safety planning so that the earnings of hazard prevention can be realized.

We also must address the questions of total cost to the community of workplace deaths and injuries, and who must be held accountable. The day and age of passing these costs onto the taxpayer through Social Security disability payments of $7 billion a year, rising welfare costs, and food stamps is over. Taxpayers no longer want to be the patsy. Economic considerations are forcing the targeting of those who fail to develop safety planning to reduce hazards in their workplaces; so the offenders' injury costs will force them to consider safety or to go out of business. This approach will enhance the nation's construction productivity and profitability.

It is a myth that the cost of injury, death, or property damage that plagues construction projects can be neutralized by insurance. There are several key issues here:

1. How much will the cost of insurance add to the cost of the construction project?
2. Will the insurance actually cover the actual losses?
3. Will the insurance company find a loophole to avoid payment?
4. Does the insurance company provide guidance on specific hazard prevention measures that will reduce the premium?

A cost analysis of hazard prevention measures also should take into consideration the cost benefits to be gained by the increased productivity that arises when people do not have to cope with workplace hazards. For example, it has been proved that staging on slipforms, which removes the hazard of falls from an elevation, creates substantial labor savings when compared to relying on workers' use of safety belts, which is not always consistent. A four-year study completed in 1957 found that men work faster when protected by scaffolds.[1] On one project, when carpenters working from whalers and without staging had to remove or replace connecting "she-bolts" to hold formwork in place, they resembled a crew of one-armed carpenters. For every dollar spent on scaffolds to protect them, the labor costs were reduced $3.00 to $6.00, which is a real return on an investment. In recommending hazard prevention measures, the quickest way to gain their acceptance is to show an increase in productivity, which means big bucks. The humanitarian aspect is a bonus. When hazards are removed from construction sites, workers can concentrate on the work at hand, as they do not have to work around a set of hazardous circumstances. The formula is: "Remove hazards with safety planning and increase productivity."[1]

The management systems we once knew have lost their way in the new market economy. Most significant is the abandonment of middle management. In its place are many small, well-informed work teams that have the authority to solve their own problems, to take the necessary action, and to move quickly on. The conventional practice of advising the foreman, who in turn must tell the supervisor, and so on up to the top with no independent action permissible, is out of touch with today's world. The question is: "How can the construction manager and project safety engineer make safety planning work without middle management?" That is easy! The use of a viable construction safety plan will inform work teams of the hazards they will encounter and tell how they can be overcome. When circumstances not identified in the plan arise, the team will know that something is amiss and that action must be taken immediately at their level to make the needed adjustments. In the past, workers always thought they had to "cope" with hazardous circumstances. As stated in earlier chapters, a rapid response to hazardous circumstances is a basic requirement of safety planning. For example, consider the case where powerlines have not been relocated as required in the safety plan by the time a crane and its operator and rigging crew arrive. The crane operator and the rigging crew must do three things according to the plan: (1) notify the electric utility; (2) notify the project safety engineer; and (3) map the danger zone (as outlined in Chapters 11 and 14), using the crane only on work that can be done outside the danger zone and where the boom cannot penetrate into the danger zone from any position. (See Figures 14-6 and 14-7.)

The new economic incentives for hazard prevention do not allow middle managers to justify their boxes on the organization chart by being approval or disapproval gates. Today's safety engineer must provide the safety coordination necessary to help individual work teams overcome hazardous circumstances as they arise; the engineer must give an immediate networking response when informed of a change of circumstances and should initiate alternative actions —such as immediately requesting a utility to deenergize powerlines, or rescheduling crane use until the powerlines have been relocated, or finding other crane use tasks that can be accomplished safely by the crane and the rigging crew. The lesson in this example is that under no circumstances should crane operations be commenced adjacent to energized powerlines, and those responsible for defaulting on the safety planning schedule will be held accountable for doing so. Two benefits result from such prompt operational follow-up: first, a high potential for injury from an armed hazard is avoided; and second, the staff members realize they are in for a real hassle when they ignore the safety planning schedule.

In the past many safety considerations got lost in the bureaucratic hierarchy. Now, with safety planning, a road map is available that gives workers the authority to control hazards at the operating level. The old concept of waiting until the safety inspector found the hazard allowed middle managers and work crews to ignore safety, and was an invitation to catastrophe. Phase safety planning that includes worker input and training develops highly productive individual work teams with little exposure to hazards. The safety engineer who has access to a safety information highway is ahead of the game by knowing the hazards that must be considered in safety planning and worker training. Today's safety engineers cannot wait until insurance companies and others provide needed hazard information, but must develop their own contacts and information resources.

We will have a free marketplace for safety, and hazard prevention will have a true economic value when the laws are changed that grant immunity to employers who deliberately provide an unsafe workplace, and when insurance companies collectively recognize that they are courting economic disaster when they do not require safety planning to be in place before coverage is written.

At first glance, this chapter may seem to speak only to the big contractor on a large project, but there are many economic incentives for the small subcontractor. Small subcontractors who are not big enough to have even a part-time safety engineer on the staff can be their own construction safety managers and can initiate a "mentor" program that relies upon the voluntary assistance of the general or the prime contractor's safety engineer for general guidance. Another approach is to explore the expertise available among retired safety engineers, who may be available for consultation on the development of a safety plan for a high-risk activity. Retired mentors usually know where to look for reference information for the control of specific hazards. The safety engineering profession has expanded rapidly in the last three decades, when many have entered the profession in their early forties. Only recently have a significant number of new entries come into the field with college training. During the 1990s and the first decade of the twenty-first century many safety professionals will be retiring, who will have valuable expertise in construction safety and should be encouraged to serve as consulting mentors. The fees they charge will be returned a hundredfold by the reduction of injury and property damage costs.

The bottom line for reaping the economic benefits of safety planning is to give workers at the operating level the same information that the boss gets or should be getting. They will respond with good ways for making a safer workplace if given hazard information and proper training. Construction managements that do not share the hazard information in their safety plan with the operating level will find themselves at an economic dead end, as they will no longer be able to compete with those in companies that have controlled their hazards through safety planning and the sharing of information with the workers who will be exposed to the hazards. Safety is everyone's business, but it must start at the top with safety planning.

REFERENCE

1. MacCollum, David V., "How scaffolding cuts costs," *Western Construction,* Sept. 1957.

# Chapter *33*

## *Training the Operating Staff of the Completed Facility*

We trained hard . . . but it seemed that every time we were beginning to form up into teams we would be reorganized. I was to learn in life that we tend to meet any new situation by reorganization; and a wonderful method it can be for creating the illusion of progress while producing confusion, inefficiency and demoralization.

(Petronius Arbiter, first century A.D.)

A new plant is useless without a well-trained staff to operate it. A complete construction safety plan must include a plan for the number of people required to operate the facility, the qualifications they must meet, and the training they will need. In many instances the manufacturer of specialized equipment can provide operating manuals and specify the level of education and experience required of applicants. Manufacturers' representatives often are the persons most qualified to conduct preliminary classroom training and monitor hands-on operation.

The objectives of such planning for operating-staff training should be:

1. To reduce the time period between facility completion and actual full production.
2. To avoid injury to operating personnel and others.
3. To avoid equipment and facility damage due to inexperience.
4. To avoid production losses due to inexperience.

The scheduling of such training should be included in the Critical Path Method (CPM) perk charts (as discussed in Chapter 11).

Training should include the opportunity for operators to witness engineering performance tests in which it is determined whether the equipment actually performs correctly as specified. Often new operators can assist in start-up testing to determine if any hazardous circumstances are apparent to them (as discussed in Chapter 26). Close coordination of training and testing gives operating personnel valuable insight into the initial functioning of the mechanical components of the facility, especially while manufacturers' representatives are available to oversee start-up activities. The value of such coordination lies in the one-on-one contact of the operating staff with the manufacturers' representatives, who can explain and remedy any discrepancies that may arise and can provide valuable insight for future operations.

An excellent learning method is to have each student compile a step-by-step checklist for starting up, operating, and stopping each component. This checklist should include references to pertinent sections of the manufacturer's manual. Such a checklist soon reveals any omissions in sequential operating steps. Where specific omissions or variations in instrument readings are found, any special comments of the manufacturer's representative need to be included in the checklist for future reference. Having manufacturers' representatives present during operating-staff training will give the needed depth to the learning experience.

The inclusion of start-up procedures as part of training also makes future operators aware of installation glitches or errors that require fine-tuning or adjustment. Possible failure modes of the equipment must be clearly identified in this training so that operators will know when safeguards and warnings are inadequate to overcome hazardous circumstances, and they can ask that safety measures be installed and posted.

An excellent time to have a system safety hazard analysis conducted by a system safety engineer to identify likely failure modes of complex systems is during this training. A system safety analysis can add immeasurably to the quality of the training, as it usually stimulates student participation in uncovering error-provoking circumstances that arise in operating particular components and in their interface with other components. Such an analysis also enhances the start-up test by providing an independent review that may identify hazards overlooked at the time of design, and it makes the operating staff more aware of hazards and their own limitations when using the equipment.

# Chapter 34

## *Fire Prevention Planning*

If a man starts a fire in his field and it spreads through the weeds to another man's field and burns up grain that is growing or has been cut and stacked, the one who started the fire is to pay for the damages.

(Exodus 22:6)

F ire prevention is an ongoing activity for the entire time of construction. The following fire prevention guidelines have been listed to assist in all facets of construction planning.

### GENERAL REQUIREMENTS

1. A monthly survey of the suitability and effectiveness of fire prevention and protection measures should be made by a qualified person, and records of the survey findings and recommendations should be retained on file by the construction manager.
2. When unusual fire hazards exist or fire emergencies develop, additional protection should be provided as required by the construction manager.
3. Fires and open flame devices should not be left unattended.
4. All sources of ignition should be prohibited within fifty feet of operations with a potential fire hazard. Smoking should be prohibited in all areas

where flammable, combustible, or oxidizing materials are stored. The areas should be conspicuously and legibly posted with signs: "NO SMOKING OR OPEN FLAME."

5. Underground areas where a danger of fire exists should not be used for storage of flammable or combustible materials.
6. Noncompatible materials that may create a fire hazard should be segregated by a barrier having a fire resistance of at least one hour.
7. Paint-soiled clothing and drop cloths, when not in use, should be stored in well-ventilated steel cabinets or containers.
8. Insulating material with a combustible vapor barrier should be stored at least twenty-five feet from buildings or structures, and only the quantity required for one day's use should be permitted in buildings under construction.
9. Disposal of combustible waste materials should be in compliance with applicable fire and environmental laws and regulations. Oil-soaked materials and paint-saturated debris should be removed from the premises on a daily basis.
10. When outside help is relied upon for fire protection, a written agreement should be made with appropriate parties, stating the terms of the arrangement and details for fire protection services.

## BURNING REQUIREMENTS

1. Burning should be done in accordance with EPA requirements. Burning areas should be established by the construction manager in coordination with those responsible for monitoring fire potential at the location of the proposed burning area.
2. Burning operations should comply with federal, state, and local regulations and guidelines.
3. A force sufficient to control and patrol burning operations should be maintained until the last embers have been extinguished.
4. Bump blocks should be provided for trucks backing into a fire or a burning pit.

## BUILDING REQUIREMENTS

1. In permanent buildings, low-density fiberboard, combustible insulation, or a vapor barrier with a flame spread rating greater than 25 should not be installed.

2. Temporary enclosures should be covered with flameproof tarpaulins or material of equivalent fire-resistance characteristics.

3. The minimum space between one-story non-fire-resistive temporary buildings should be twenty feet, and between two-story non-fire-resistive buildings twenty-five feet; and other fire-resistive buildings should comply with recommendations of the National Fire Protection Association (NFPA).

4. All small sheds should be separated by at least ten feet on each side from each other and from other buildings.

5. Fire lanes to provide access to all areas should be established and maintained free of obstruction.

6. "No parking" areas should be designated around fire hydrants and firefighting equipment.

7. Electrical lighting should be the only means of artificial illumination in areas where flammable liquids, vapors, fumes, dust, or gases are present; all electrical equipment and installations in hazardous locations should be in accordance with the *National Electrical Code* for hazardous locations; and globes or lamps should not be removed or replaced and repairs should not be made on the electrical circuit until it has been deenergized.

8. Clearance should be maintained around lights and heating units to prevent ignition of combustible materials.

9. Water supply and distribution facilities for fire fighting should be provided and maintained in accordance with recommendations of the NFPA or U.S. Coast Guard regulations.

## FIRE PROTECTION DURING CONSTRUCTION

1. All combustibles should be shielded from flames of torches and welding arcs.

2. Formwork and scaffolding should be protected from exposure to and spread of fire.

3. Fire cutoffs should be retained in buildings undergoing alterations or demolition until operations necessitate their removal.

4. Where a water distribution system is required for the protection of buildings or other structures, water mains and hydrants should be installed prior to or concurrently with the construction of the facilities. A temporary system equivalent to a permanent system should be provided until the latter is in operation.

5. As soon as possible, operable permanent (fixed) fire-extinguishing equipment and a water supply should be installed for fire protection. A sprinkler system should be installed soon after building construction and placed in operation upon the completion of each story as soon as the law permits its use.

6. During demolition or alterations, existing automatic sprinkler systems should be retained in service as long as is reasonable. The modification of sprinkler systems to permit alterations or additional demolition should be expedited so that the system may be returned to service as quickly as possible. Sprinkler control valves should be checked daily at the close of work to ascertain that the protection is in service. The operation of sprinkler control valves should be permitted only when approved by the construction manager.

7. Construction of fire walls and exit stairways required for completed buildings should have priority; fire doors, with automatic closing devices, should be hung on openings as soon as they are practical.

## FLAMMABLE AND COMBUSTIBLE LIQUIDS

1. A qualified person should supervise the storage, handling, and use of flammable and combustible liquids.

2. "NO SMOKING OR OPEN FLAME" signs should be posted in all areas where flammable and combustible liquids are stored, handled, and processed.

3. At least one portable fire extinguisher rated not less than 20-B should be located outside of, but not more than ten feet from, the door opening into any room used for the storage of more than sixty gallons of flammable or combustible liquids.

4. At least one portable fire extinguisher having a rating of not less than 20-B should be located not less than twenty-five or more than seventy-five feet from any outside flammable liquid storage area.

5. At least one portable fire extinguisher rated 20-B:C should be provided on all tank trucks or other vehicles used for transporting and/or dispensing flammable or combustible liquids.

6. Each service or refueling area should be provided with at least one fire extinguisher rated not less than 20-B:C and located within seventy-five feet of each pump, dispenser, underground fill-pipe opening, and lubrication or service area.

7. When not in use, flammable liquids should be kept in closed containers.

8. Workers should avoid getting their clothing contaminated with flammable or combustible fluids. If their clothing does become contaminated, they should not be allowed to continue work and must remove or wet down the clothing as soon as possible.

9. No flammable liquid with a flash point (closed cup test) below 100°F should be used for cleaning purposes or to start or rekindle fires.

10. In all areas where flammable and combustible liquids are handled or used, adequate ventilation should be provided to prevent hazardous levels of flammable vapors from accumulating.

11. In areas where flammable and combustible liquids are handled or stored, a self-closing metal refuse can, listed by a nationally recognized testing laboratory, should be provided and maintained in good condition.

12. During handling of flammable and combustible liquids, workers should use only flashlights and electric lanterns listed by a nationally recognized testing laboratory for use in such hazardous areas.

13. For storage or handling of flammable and combustible liquids, workers should use only tanks, containers, and pumping equipment, whether portable or stationary, listed by a nationally recognized testing laboratory or approved by, and labeled or tagged in accordance with, the federal agency having jurisdiction over them.

14. Unopened containers of paints, varnishes, lacquers, thinners, solvents, and other flammable and combustible liquids should be kept in a well-ventilated location, free of excessive heat, smoke, sparks, flame, or direct rays of the sun.

15. Tightly capped safety containers with flame arresters should be used for hand-handling of all flammable and combustible liquids. Containers for flammable liquids having a flash point at or below 73°F should be painted red with a yellow band around the can and the name of the contents printed on the container. For quantities of one gallon or less, the original container may be used.

16. Entrances to buildings in which flammable or combustible liquids are stored or processed must be kept locked when not occupied, and only authorized persons should be permitted to enter such buildings.

17. Flammable and combustible liquids should not be stored in areas used for exits, stairways, or passageways.

18. Materials that react with water and create a fire hazard should not be stored in the same area with flammable or combustible liquids.

19. No more than twenty-five gallons of flammable or combustible liquid should be stored in a room outside of an approved metal storage cabinet. For quantities greater than twenty-five gallons, an approved metal storage

cabinet that meets the requirements of NFPA 30 should be used for indoor storage and labeled in conspicuous lettering "FLAMMABLE—KEEP FIRE AWAY." No more than 60 gallons of flammable or 120 gallons of combustible liquid should be stored in any one storage cabinet; and no more than three such cabinets should be located in a single storage area.

20. Oil, grease, or fuel stored underground should be kept in tightly sealed containers in fire-resistant areas and at safe distances from explosive magazines, electrical installations, and shaft stations.

21. Indoor storage rooms for flammable and combustible liquids should be constructed to meet NFPA fire-resistance requirements.

22. When indoor storage rooms have automatic extinguishing systems, the openings to other rooms should have noncombustible liquid-tight raised sills or ramps at least four inches in height, and the floor of the storage room should be at least four inches below the surrounding floor, or the storage room should have an open-grated trench that drains to a safe location. The openings should have approved self-closing fire doors, and the room should be liquid-tight where the walls join the floor. Wood of at least one inch nominal thickness may be used for shelving, racks, dunnage, floor overlay, and so forth.

23. When containers of flammable or combustible liquids over thirty gallons in capacity are stored indoors, they should not be stacked one upon the other.

24. In every indoor storage room, one clear aisle at least three feet wide should be maintained.

25. Indoor storage rooms should be ventilated by a gravity or mechanical exhaust system that commences not more than twelve inches above the floor and is designed to provide for a complete change of air within the room at least six times per hour. A mechanical exhaust system should be controlled by a switch located outside the door, with the ventilating equipment and any light fixtures operated by the same switch. An electric pilot light should be installed adjacent to and operated by the same switch if flammable liquids are dispensed within the room. If a gravity ventilation system is used, the fresh air intake as well as the exhaust outlet from the room should be on the exterior of the building in which the room is located.

26. Flammable and combustible liquids in quantities greater than the amounts required for one day's use should not be stored in buildings under construction.

## Outside Storage of Flammable and Combustible Liquids

1. Storage containers (not more than 60 gallons each) should not exceed 1,100 gallons in any one group.

2. Groups of containers should be separated by a five-foot clearance.
3. Groups of containers should not be closer than fifty feet to a building.
4. Outdoor portable tanks should be at least fifty feet from buildings.
5. Portable tanks not exceeding 660 gallons in capacity should be provided with emergency venting and other devices as specified in NFPA 30.
6. Storage areas should be kept free of accumulation of weeds, debris, and other combustible materials not necessary to the storage.

## STORAGE TANKS

1. Storage tanks should be equipped with relief vents.
2. Tank vents should not be located close to open flames, stacks, heating apparatus, or any other source of ignition.
3. Water drawoff valves should be of the antifreeze type or insulated to prevent freezing.
4. Aboveground storage tanks should be diked or curbed, or other suitable means provided to prevent the spread of liquids in case of leakage in tanks or piping. The diked or curbed area should have a capacity equal in volume to at least that of the largest tank plus 10 percent of all other tanks enclosed. Provision should be made for draining off accumulations of ground- or rainwater or spills. Drain plugs should remain in place except during draining. Drains should terminate at a safe location and should be accessible to operation under fire conditions.
5. Underground tanks used during construction should not be abandoned. They should be either converted for permanent use consistent with EPA regulations or removed. Appropriate soil tests should be made to ascertain whether surrounding soil was contaminated.

## DISPENSING OF FLAMMABLE OR COMBUSTIBLE LIQUIDS

1. Dispensing outlets for aboveground tanks should be equipped with a nationally listed automatic-closing valve, without a latch-open device.
2. Flammable-liquid-dispensing systems should be electrically bonded and grounded. All tanks, hoses, and containers of five gallons or less should be kept in metallic contact while flammable liquids are being transferred;

the transfer of flammable liquids in containers in excess of five gallons should be done only when the containers are electrically bonded.

3. Within a building or outside, flammable or combustible liquids should be drawn from, or transferred into, vessels, containers, or tanks only through a closed piping system; from safety cans, by means of a device drawing through the top; or from a container or portable tanks, by gravity or pump, through an approved self-closing valve. Transfer by means of air pressure on the container or portable tanks is prohibited.

4. Areas in which flammable or combustible liquids are transferred in quantities greater than five gallons from one tank or container to another should be separated from other operations by at least twenty-five feet or by a barrier having a fire resistance of at least one hour. Drainage or other means should be provided to control spills. Natural or mechanical ventilation should be provided to maintain the concentration of flammable vapor at or below 10 percent of the lower flammable limit.

5. Dispensing units should be protected against collision damage.

6. Dispensing nozzles and devices for flammable liquids should be of an approved type.

7. Lamps, lanterns, heating devices, and similar equipment should not be filled while hot and should be filled only in well-ventilated rooms free of open flames or in open air. They should not be filled in storage buildings.

## SERVICE AND REFUELING AREAS

1. Dispensing hoses should be of an approved type. Dispensing nozzles should be of an approved automatic-closing type without a latch-open device.

2. Equipment using flammable liquid fuel should be shut down during refueling, servicing, or maintenance. This requirement may be waived for diesel-fueled equipment serviced by a closed system with attachments designed to prevent spillage.

3. Dispensing of flammable fluids from tanks of fifty-five gallon or greater capacity should be done by an approved pumping arrangement. Transfer by air pressure on a container or a portable tank is prohibited.

4. Clearly identified and easily accessible switch(es) should be provided at a location remote from dispensing devices to shut off the power to all dispensing devices in an emergency.

## TANK CARS OR TRUCKS

1. Tank cars or trucks should be spotted and not loaded or unloaded until the brakes have been set and the wheels chocked.
2. Tank cars or trucks should be attended for the entire time that they are being loaded or unloaded. Precautions should be taken against fire or other hazards.
3. Tank cars or trucks should be properly bonded and grounded while being loaded or unloaded. Bonding and grounding connections should be made before dome covers are removed, and should not be disconnected until such covers have been replaced. Internal vapor pressure should be relieved before dome covers are opened.

## LIQUEFIED PETROLEUM GAS (LP-GAS)

1. Storage, handling, installation, and use of LP-gases and systems should be in accordance with NFPA Standard 58 and regulations of the U.S. Coast Guard, as applicable.
2. LP-gas containers, valves, connectors, manifold-valve assemblies, regulators, and appliances should be of an approved type.
3. Any appliance that was originally manufactured for operation with a gaseous fuel other than LP-gas and is in good condition may be used with LP-gas only after it is properly converted, adapted, and tested for performance with LP-gas.
4. Polyvinyl chloride and aluminum tubing should not be used in LP-gas systems.
5. Every container and vaporizer should have one or more safety relief valves or devices arranged to vent freely to the outside air and discharge at a point not less than five feet horizontally from any building opening that is below the discharge point.
6. Container safety relief devices and regulator relief vents should be located not less than five feet in any direction from air openings into sealed combustion system appliances or mechanical ventilation air intakes.
7. Shutoff valves that would disable safety relief devices should not be installed.
8. Valves, fittings, and accessories connected directly to the container, including primary shutoff valves, should have a rated working pressure of

at least 250 psi and should be of a material and design suitable for LP-gas service.

9. Connections to containers, except safety relief connections, liquid level gauging devices, and plugged openings, should have shutoff valves located as close to the container as practical.

10. Valves in the assembly of multiple container systems should be arranged so that replacement of containers can be made without shutting off the flow of gas in the system. This is not to be construed as requiring an automatic changeover device.

11. Regulators and low-pressure relief devices for multiple container systems should be rigidly attached to the cylinder valves, cylinders, supporting standards, or building walls or otherwise rigidly secured, and should be protected from the elements.

12. LP-gas containers and equipment must not be used in unventilated spaces below grade in pits, below-decks, or in other such spaces where dangerous accumulations of heavier-than-air gas may accumulate because of leaks or equipment failure.

13. Welding must be prohibited on LP-gas containers.

14. Equipment using LP-gas must be shut down during refueling operations.

15. Fuel containers of LP-gas for motor vehicles must not be filled from bulk storage containers less than ten feet from the nearest masonry-walled building, less than twenty-five feet from the nearest building of other construction, and, in any event, less than twenty-five feet from any building opening.

16. Portable containers or containers mounted on skids must not be filled from storage containers less than fifty feet from the nearest building.

## INSTALLATION, USE, and STORAGE OF LP-GAS OUTSIDE OF BUILDINGS

1. Containers should be upright upon firm foundations or otherwise firmly secured. Flexible connections (or other special fixtures) must be provided to protect against possible settlement on the outlet piping.

2. Containers should be in a suitable ventilated enclosure or otherwise protected against tampering.

3. At least one portable fire extinguisher rated no less than 20-B:C should be provided in all storage areas.
4. LP-gas must not be stored inside buildings.

## TEMPORARY HEATING DEVICES

1. Unvented gas heating devices or unvented combustion heaters must not be used.
2. Only those temporary heating devices approved by the Underwriters' Laboratories (UL) should be used. Each heater should have a safety data plate permanently affixed by the manufacturer that contains information on the following:
   a. Clearances from combustible materials.
   b. Ventilation (minimum air requirements for fuel combustion).
   c. Fuel type and input pressure.
   d. Lighting, extinguishing, and relighting.
   e. Electrical power supply characteristics.
   f. Location, moving, and handling.
   g. Name and address of the manufacturer.
2. A positive inspection program should be established for:
   a. Determining if a proper flue is in place to vent exhaust fumes.
   b. Determining that there is an adequate outside air source for the combustion chamber.
   c. Proper placement and servicing.
   d. Ensuring safe clearance from combustible material.
   e. Close surveillance.
   f. Safe fuel storage and refueling.
   g. Proper maintenance.
   h. Monitoring the airspace to see that carbon monoxide is not being generated.
3. Heater installation and maintenance should be done in accordance with the manufacturer's instructions. Heaters should be protected against damage.
4. Combustion heaters should be equipped with an approved automatic device to shut off the flow of fuel if the flame is extinguished.
5. Spark arresters should be provided on all smokestacks or burning devices

having forced drafts or short stacks permitting live sparks or hot materials to escape.

6. All gas piping, tubing, and hose should be leak-tested at normal operating pressure after assembly, by using soap suds or other noncombustible detection means (tests should not be made with a flame).

7. Fuel piping, tubing, hose, and fittings should be protected from damage and deterioration.

## FIRST RESPONSE FIRE PROTECTION

1. Portable fire extinguishers should be provided where needed, and should be inspected monthly.

2. Fire extinguishers should be approved by a nationally recognized testing laboratory and labeled to identify the listing and labeling organization and the fire test and performance standard that the fire extinguisher meets or exceeds.

3. Fire extinguishers should be marked with their letter (class of fire) and numeric (relative extinguishing effectiveness) classifications.

4. Personnel must not use fire extinguishers using carbon tetrachloride or chlorobromomethane extinguishing agents.

5. Fire extinguishers must be kept in a fully charged and operable condition and placed in a conspicuous location, distinctly marked and readily accessible.

6. No fire should be fought where the fire is in imminent danger of contact with explosives. All persons should be removed to a safe area.

7. All personnel should be trained in general principles of fire prevention, use of extinguishers, and emergency exits.

# Chapter 35

*Surviving the Crisis of Unmet Planning for*
*Changing Circumstances*

The best laid schemes of men and mice often go awry.

(Robert Burns)

T he best advice for any organization when things go wrong is not to stonewall the media. If it does, it soon will have the public reputation of not caring about the life and the safety of others.

Emergency plans first should list the phone numbers for medical assistance, fire fighters, police, maintenance people for damage control, and other emergency support services. When possible, arrangements should be made prior to construction with the closest hospital emergency room and ambulance services to make them aware of the location of the project. For remote locations, air evacuation arrangements should be made so that an appropriate emergency landing site can be agreed upon.

The next to be notified of any emergency should be the project construction manager and the project safety engineer. The construction manager should be the designated spokesperson in an emergency situation so that the media and other officials have access to someone who can speak with authority.

The following guidelines for crisis management should be included in any construction safety plan:

1. Let everyone on the project know that what happens, happens, and no one should try to cover up information. Many problems arise when there is

even suspicion of a cover-up. Cover-ups have the nasty habit of coming back and eating their perpetuators alive.

2. If the cause is not immediately apparent, say so, and do not share speculations with the media.

3. Have an investigation board composed of safety and engineering experts available on an "on-call" basis for immediate response. This board's investigation and report should be prompt and concise, and should:
   a. Identify all hazards involved.
   b. List necessary hazard prevention measures.
   c. List deviations from applicable safety requirements and/or regulations and available technology.
   d. List a bibliography of supporting references and literature that discuss the hazards and their control. Also, a summary of similar occurrences should be included in the reference data.

4. Promptly provide the press and officials with brief and concise facts about the occurrence, and follow up with the findings of the investigating board. Address the hazards that were the cause of the crisis. Never talk about who may be at fault or to blame. Emphasize preventive measures being taken to avoid further injury or damage.

A crisis can be best managed by promptly providing the investigating government agencies with a thorough analysis that clearly identifies the hazards, shortcomings, deficiencies, and violations that were the cause of the crisis. Strong emphasis should be placed upon the preventive measures that will be instituted immediately. This will reduce the negative response that so often occurs after a disaster.

The best crisis-avoidance policy is a construction safety plan. The public is always interested in large, local construction projects. One way to acquaint television and newspaper reporters and other interested officials with the project is to sponser a "media day" in which they can become familiar with the project in a "nonemergency" setting. This not only will provide the media with information about whom they should contact in the event of an emergency, but, most important, will show whom the construction manager should contact when a press release must be issued. You should treat the media as honored guests, providing a place inside the gates of the project where they can meet with the construction manager and appropriate staff. Once they are inside the premises, they should understand that they must abide by the project's safety rules. Having such a place will also keep the media away from any activists who might be causing a disturbance outside that could prevent the correct facts about the project from being presented. With members of the media on-site, organized photographic and video opportunities should be arranged so that they have an opportunity to see for themselves what the project is all about. In the event of a crisis, the media day will pay big dividends because reporters will be familiar with the project, will have met the construction manager and some of the staff, and will be receptive to the assistance provided to them. The construction manager must project a strong leadership presence by telling the reporters what has occurred and what has been done, and that as soon as factual information is available, they will be the first to know it.

The role of the project safety engineer is to provide a hazard analysis and all supporting data, literature, and pertinent requirements so that the construction manager can tell the press what the failure mode was and how it is being overcome. Excuses only create doubt, apprehension, and a loss of public confidence. By keeping the focus on physical hazards and the preventive actions being taken, a posture of positive commitment to a safe workplace will be presented. "Mickey Mouse" remedies that do not eliminate hazards must be avoided, as the public will see them as management's failure to "bite the bullet" for safety.

Uninjured witnesses of any occurrence should be located promptly and asked about what they observed. Witnesses have important information needed by the investigating board and should be individually interviewed in a quiet, comfortable location without distractions or peer or supervisory pressure. Recording of their statements must be done in a nonhostile manner to keep them from having any mental reservations or fear of future criticism. People who observed the physical conditions prior to the occurrence make good witnesses also, including the workers on a previous shift and the maintenance people. It is important to obtain the individual statements of witnesses as quickly as possible, before they have had an opportunity to discuss their observations with others, as it is human nature to accommodate the observations or viewpoints of others. Questions must never pressure witnesses for specific conclusions. What is essential is to have accurate information about what they saw. After they have made their recorded statements, they can continue to be of assistance to the investigating board by assisting in the recovery or the identification of specific items that might help in the evaluation or the analysis of the occurrence. Witnesses engaged in assisting in the investigation should not be exposed to public pressure for information, as this often leads to a record of conflicting statements. Corporate counsel and other lawyers should not be allowed to intervene or be involved in the investigative or hazard-analysis processes, as their presence tends to impede an unbiased fact-finding process. Only after the investigation is completed and the hazard analysis is made should the written report be made available as public information to the legal staff. Keeping injury or damage reports secret creates public distrust. Openness is the key to successful crisis management.

Construction safety planning is the key to successful building and will guarantee a well-coordinated project that avoids most delays and unexpected changes because of oversights and misunderstandings. Its greatest benefit is a safer workplace that is not plagued by injury, death, or other crises.

Crisis circumstances always produce stress; so clear and concise answers about previously identified hazardous conditions need to be available for the construction manager as spokesperson. To assist him/her in this, the project safety engineer should provide summaries of any similar or closely related hazards that were discussed at monthly safety oversight meetings (Chapter 19) or that surfaced during project monitoring (Chapter 20), so that the construction manager will not be caught offguard if questions are raised by the media concerning possible prior knowledge. The project safety engineer also should summarize the findings concerning a crisis circumstance so that current phase planning and safety planning for future projects can incorporate the lessons learned.

*Appendix 1*

## Selected References on Construction Safety

The following references should be available for construction safety planning.

1. Current edition (revised every year) of *Code of Federal Regulations*, Title 29 Labor, Chapter XVII, Occupational Safety and Health Administration (OSHA), Department of Labor, Parts 1910 for General Industry and 1926 for Construction, available from Superintendent of Documents, U.S. Government Printing Office, Washington, DC 20402.
2. Current edition of U.S. Army Corps of Engineers' *Safety and Health Requirements Manual*, EM 385-1-1, available from Superintendent of Documents, U.S. Government Printing Office, Washington, DC 20402.
3. Current edition of *Accident Prevention Manual for Industrial Operations*, Volume I, *Administration and Programs* and Volume II, *Engineering and Technology*, National Safety Council (NSC), 444 North Michigan Avenue, Chicago, IL 60611.
4. *Industrial Data Sheets* for construction, mining, and quarrying, National Safety Council, 444 North Michigan Avenue, Chicago, IL 60611, as follows:

| | |
|---|---|
| Acetylene | No. 494 |
| Aerial Baskets | 572 |

5. Current standards of the American National Standards Institute (ANSI), 1430 Broadway, New York, NY 10018, most generally relied upon for construction safety:

A10.3  Powder-Actuated Fastening Systems

A10.4  Personnel Hoists and Employee Elevators

A10.5  Material Hoists

A10.6  Demolition

A10.7  Commercial Explosives and Blasting Agents—Safety Requirements for Transportation, Storage, Handling and Use

A10.8  Scaffolding

A10.9  Concrete Construction and Masonry Work

A10.10 Temporary and Portable Space Heating Devices and Equipment Used in the Construction Industry

A10.11 Safety Nets Used during Construction, Repair and Demolition Operations

A10.12 Excavation (under development)

A10.13 Steel Erection

A10.14 Safety Belts, Harnesses, Lanyards and Lifelines

A10.15 Dredging

A10.16 Tunnels, Shafts and Caissons

A10.17 Asphalt Pavement Construction (under development)

A10.18 Temporary Floor and Wall Openings, Flat Roofs, Stairs, Railings and Toeboards

A10.19 Pile Driving (under development)

A10.20 Ceramic Tile, Terrazzo and Marble Work

A10.22 Rope-Guided and Nonguided Workmen's Hoists

A10.24 Roofing (under development)

A10.27 Asphalt Mixing Plants for Construction Projects (under development)

A10.28 Work Platforms Suspended from Cranes or Derricks

A10.30 Drilled Caissons (under development)

A10.31 Digger Derricks

A10.32 Fall Prevention Systems (under development)

A10.33 Safety and Health Program Requirements for Multi-employer Projects

A10.34 Public Protection (under development)

A10.35 High-Pressure Hydro Blasting (under development)

A10.36 Dry Diamond Saws (under development)

A10.37 Debris Nets (under development)

A10.38 Special Safety Programs

A10.40 Duties and Qualifications of Individuals Assigned Construction Safety and Health Responsibilities

A14.1 Portable Wood Ladders Safety Requirements

A14.2 Portable Metal Ladders Safety Requirements

A14.4 Job-Made Ladders Safety Requirements

A14.5 Portland Plastic Reinforced Ladders Safety Requirements

A64.1 Requirements for Fixed Industrial Stairs

A90.1 Safety Standard for Manlifts

A92.1 Manually-Propelled Mobile Ladder Stands and Towers

A92.2 Vehicle-Mounted Elevating and Rotating Aerial Devices

A92.3 Manually-Propelled Elevating Work Platforms

A92.5 Self-Propelled Boom-Supported Elevated Work Platforms

A92.6 Self-Propelled Elevating Work Platforms

A120.1 Safety Code for Powered Platforms for Exterior Building Maintenance

B30.1 Jacks

B30.5 Mobile and Locomotive Cranes

B30.6 Derricks

B30.7 Base Mounted Drum Hoists

B30.8 Floating Cranes and Floating Derricks

B30.9 Slings

B30.10 Hooks

B30.12 Handling Loads Suspended from Rotercraft

B30.14 Side Boom Tractors

B30.15 Mobile Hydraulic Cranes

B30.19 Cableways

B30.22 Articulating Boom Cranes

C2        National Electrical Safety Code

D6.1     Manual on Uniform Traffic Control Devices

6. Current edition of *SAE Handbook*, Volume 4, *On-Highway Vehicles and Off-Highway Machinery*, Society of Automotive Engineers, Inc., 400 Commonwealth Drive, Warrendale, PA 15096.

7. *Safety and Health for Engineers*, Roger L. Brauer, Van Nostrand Reinhold, 115 Fifth Avenue, New York, NY 10003.

8. *Dangerous Properties of Industrial Materials*, N. Irving Sax, Van Nostrand Reinhold, New York.

9. *Crane Hazards and Their Prevention*, David V. MacCollum, American Society of Safety Engineers, 1800 East Oakton Street, Des Plaines, IL 60018.

10. *Crane Handbook*, Donald Dickie, Construction Safety Association of Ontario, Canada, 74 Victoria Street, Toronto, Ontario, Canada M5C2A5.

11. *Basic Guide to System Safety*, Jeffrey W. Vincoli, Van Nostrand Reinhold, New York.

12. *Basic Guide to Environmental Compliance*, Jeffrey W. Vincoli, Van Nostrand Reinhold, New York.

13. *Basic Guide to Accident Investigation and Loss Control*, Jeffrey W. Vincoli, Van Nostrand Reinhold, New York.

14. *Introduction to Fall Protection*, J. Nigel Ellis, American Society of Safety Engineers, Des Plaines, IL.

15. *Federal Register*, Tues. Aug. 9, 1994, 29 CFR, Parts 1910 and 1926, "Safety Standards for Fall Protection in the Construction Industry; Final Rule," with supporting data submitted by the Assistant Secretary of Labor Joseph A. Deer, July 29, 1994.

# Critical Lift Plan for Safe Crane Use

This appendix contains questions and considerations for safe crane use.

## PART I. LOCATION WHERE LIFT IS TO BE MADE

1. Is there safe access for the crane?
2. Are there powerlines that the crane could reach? (See Chapter 14, Figures 14-4, 14-5, and 14-6.)
3. Are there trees, buildings, or other obstructions that must be avoided?
4. Does the location call for blind lifts requiring a communication system?
5. Is there adequate safe working and ground space for the riggers and others, for the outriggers, and so on?
6. Are the ground conditions adequate to support the crane, or is matting required?
7. Is the ground level, or is fill or cribbing necessary?
8. What are the anticipated weather conditions? Has lightning, high winds, rain, snow, or ice been forecast?
9. Will the operator or the signaler have to look into the sun?
10. Does the lift have to be made at night, what lighting needs to be provided, and is it glare-free?

11. Will the work location be next to water—on embankments, docks, bridges, or a barge—which requires special water safety provisions consistent with U.S. Coast Guard requirements, such as life preservers, emergency boats, and so forth?
12. Has safe access been provided for bringing the loads to the site?
13. Does the load have to be lifted over buildings where people may be inside, or in other high-risk locations?
14. Can the crane be isolated from pedestrian and vehicular traffic?
15. In the event that an emergency occurs, is there a place where the load can be lowered or dropped if the crane starts to tip or fail?
16. Will a pinchpoint be created between the rotating crane cab/counterweight and the building, wall, poles, or other fixed object? If so, what barricades or warnings will be provided?

## PART II. THE LOAD TO BE LIFTED

1. How heavy is the load, and will it require a special crane or multiple cranes?
2. How will the weight of the load be verified, as shipping weights are often erroneous?
3. What are the physical dimensions and the shape of the load?
4. Where is the center of gravity of the load?
5. Does it have lifting lugs? Is the positioning of straps, cables, or chains specified or discretionary?
6. Is a lifting beam, spreader bar, or other special rigging required?
7. Is the load symmetrical or nonsymmetrical? What type of rigging is required?
8. From what position will the load have to be lifted?
9. Does the load have to be rotated or have a change of axis in the lift?
10. What will be the final resting position of the load?
11. How many times does the lift have to be made?
12. Have any time limits been established for completing the lift or lifts?
13. Is the lift going to be complicated or critical and require computer-assisted analysis (CAD)?
14. Will the load be freely suspended at all times? A vibratory piledriver and puller is not a freely suspended load and should not be used without the crane manufacturer's authorization.

## PART III. THE CRANE ITSELF

1. When was the crane last inspected, load-tested, and certified? Will it need to be reevaluated?
2. Is the crane of adequate capacity? Will it be well within its rated capacity at the most critical point of the lift?
3. Is it equipped with the following functional devices?
   a. Load-moment device or load indicator?
   b. Anti-two-blocking device on both the main hoist line and the jib?
   c. Power lowering?
   d. Safety latches on the hooks?
   e. Spirit level?
   f. Boom-angle indicator?
   g. On latticework booms, boom stops or snubs and a boom-hoist limit switch?
   h. Insulated link?
   i. Proximity alarm?
4. Does the crane have a dangerously close pinchpoint (fourteen inches or less) between the rotating cab/counterweight and the carrier frame or crawler treads? If so, what barricades or warning need to be provided?

## PART IV. PERSONNEL INVOLVED

1. Who was the individual who inspected and certified the crane, and was he/she competent and qualified to make a thorough examination?
2. Who will be in charge of the lifting activity? What are his/her qualifications and experience, and does he/she have authority to oversee the use of the crane?
3. Who is the crane operator? Find out that individual's:
   a. Duties, responsibilities, and authority.
   b. Education level.
   c. Apprenticeship training and work experience by type of crane. Specify the crane's:
      (1) Capacity.
      (2) Classification (hydraulic, latticework boom, bridge, etc.).
      (3) Activity (hoisting, dragline, pile driving, etc.).
      (4) Boom length.

(5) Type (crawler, rough-terrain, tower, etc.).

d. Classroom training on crane safety.

e. Knowledge of crane safety requirements and references.

f. Physical qualifications:

(1) Age (maturity and intelligence required).

(2) Vision. ANSI/ASME B30.5, "Mobile and Locomotive Cranes," Section 5-3.1.2(b)(1) requires: "Vision of at least 20/30 Snellen in one eye and 20/50 in the other, with or without corrective lenses," and Section 5-3.1.2(b)(2) requires: "Ability to distinguish colors, regardless of position, if color differentiation is required for operation. . . ."

(3) Hearing. ANSI/ASME.5, "Mobile and Locomotive Cranes," Section 5-3.1.2(b)(3) requires: "Adequate hearing with or without hearing aid, for the specific operation. . . ."

(4) Good physical stamina.

(5) Good coordination, reaction, and tested skill level.

(6) No history of heart problems or other ailments that could produce seizures.

(7) Emotional stability.

(8) Absence of drug and/or alcohol abuse and addictions, and so forth.

g. Record of injury of property damage with operating cranes.

h. Skill. The State of New York, in licensing crane operators, uses videotapes to verify an operator's judged ability to perform the State's required operating procedures.

i. Special training.

j. Disciplinary actions for improper operations.

k. Log of operating time (when and on what equipment).

l. Actual knowledge of crane use and hazards.

m. Language skill. Special bilingual requirements may be necessary where members of minority groups do not speak English.

4. Crane operators should pass a written examination demonstrating their knowledge of the following:

a. Responsibilities of the operator, rigger, signal persons, and lift supervisor.

b. The crane manufacturer's safety requirements and the content of the crane's operator manual.

c. The crane's configuration, and being able to compute the size and shape of loads and to use the load chart to determine the crane's capacity.

d. Use and limitations of crane operator aids.

    e. Inspection, testing, and maintenance requirements.

    f. Determination of ground conditions and outrigger and matting requirements.

    g. Crane setup, assembly, dismantling, and demobilization procedures.

    h. Requirements for clearance from power sources.

    i. Signaling and communication procedures.

    j. Factors that reduce rated capacity.

5. Crane operators should pass a practical operating examination demonstrating their ability to do the following:

    a. Inspect the crane.

    b. Establish a stable foundation and level the crane.

# Index

CPSIA information can be obtained at www.ICGtesting.com
Printed in the USA

266963BV00002B/3/P

9 780471 286691